Human Interaction With Reused Soil: An Information Search

I0487749

Final Report

Manuscript Completed: November 2001
Date Published: January 2002

Prepared by
S. McCarthy, USDA/ARS/NAL
T. Nicholson, J. Philip, USNRC/RES
E. Brummett, F. Cardile, G. Gnugnoli, A. Huffert, USNRC/NMSS

U.S. Department of Agriculture
Agricultural Research Service
National Agricultural Library
10301 Baltimore Avenue
Beltsville, MD 20705-2351

T. Nicholson, NRC Project Manager

Prepared for
Division of Systems Analysis and Regulatory Effectivenesss
Office of Nuclear Regulatory Research
U.S. Nuclear Regulatory Commission
Washington, DC 20555-0001
NRC Job Code Y6227

ABSTRACT

This NUREG-series publication reports the results of a 2-year investigation to compile information intended to support the formulation and characterization of scenarios related to exposure to residual radioactivity in reused soils. This information search focused on human interactions with reused soils in the United States. Using this information, the staff and contractors of the U.S. Nuclear Regulatory Commission (NRC) are working to define realistic soil reuse scenarios and to estimate parameters for simulating exposure pathways if soil is removed from NRC-licensed facilities. NRC staff and researchers from the National Agricultural Library (NAL) conducted this investigation in two phases. Phase 1 was a general information search structured to query literature from a wide range of published scientific and trade sources. Phase 2 was a focused information search on specific parameters such as contact time, dust exposures and tillage depths identified in Phase 1. NAL staff searched additional sources and contacted individuals in the Government, academia, and commerce. This report compiles, for the first time, data and information sources for parameters specific to soil reuse. This report also provides information that is relevant for generic, as well as site-specific dose assessments, and presents typical information that may be used in future dose modeling analyses. This investigation was coordinated with a companion effort by the Federal Interagency Steering Committee on Radiation Standards on disposition of sewage sludge.

CONTENTS

CONTENTS (Continued)

CONTENTS (Continued)

APPENDICES

CONTENTS (Continued)

FIGURES

TABLES

EXECUTIVE SUMMARY

This NUREG-series publication reports the results of a 2-year information search focused on human interactions with reused soils in the United Sates. This investigation sought to compile information intended to support the formulation and characterization of scenarios related to exposure to residual radioactivity in reused soils. Developed under an interagency agreement between the U.S. Nuclear Regulatory Commission (NRC) and the National Agricultural Library (NAL) of the U.S. Department of Agriculture's Agricultural Research Service, this report documents the process and results of an extensive literature review and focused information search on human contact with soil. This report was issued in June 2000 as Draft NUREG-1725, "Human Interaction with Reused Soil: A Literature Search," to solicit public comments and additional information.

This report describes the methodologies used in developing the information search strategies and in selecting the database sources. These strategies and sources were reviewed by external reviewers, whose comments are included in the report. The primary focus of the work was to identify documented, verifiable references for the NRC staff to use modeling dose exposures related to soil reuse.

This report describes the results of the literature search pertaining to human interactions with reused soil, and presents typical information resulting from the search that might be used in future modeling. This report provides information on the ways in which soils are reused in commerce (e.g., landscaping) or by the general public in the United States. This information is an important part of the technical basis for assessing the possible exposures that could result if soil is released from NRC-licensed facilities. Specifically, this information will be used to develop soil reuse scenarios for use in dose assessments. To develop these scenarios, it is necessary to obtain data on the forms of human contact with soils in the United States. Using this information, the NRC staff and contractors are working to define realistic soil reuse scenarios and to estimate parameters for simulating exposure pathways involving soil removed from NRC-licensed facilities.

NRC staff and researchers from the NAL conducted this investigation in two phases. For the initial literature search in Phase 1, the information sources searched included the Dialog® (an online system of more than 500 databases), the Internet, and other sources. With support from NRC staff and external reviewers (non-NAL library and information science and soil science professionals), the NAL researchers developed targeted search strategies to retrieve relevant items from the Dialog® databases. These search strategies were structured to query literature from science publishers, academic presses, professional societies, trade journals and bulletins, theses and dissertations, as well as information published in industry standards, newspapers, company reports, statistical sources, etc. The Dialog® search was developed by creating three main categories or concept sets, including "General" (actions or activities of humans with soils), "Particular" (specific identified types of human-soil interactions), and "Volume" (volume, quantity, or economic terms that quantify or delineate the extent of human contact with soil).

In the Phase 1 effort, the researchers screened the initial records found in the Dialog® database searches for relevance to the scope of this research effort. Through the detailed review of these records by NAL and NRC researchers, a small number of documents were identified as relevant.

In the Phase 2 effort, the emphasis was on identifying information sources for formulating and estimating parameters to be used in modeling soil reuse scenarios involving potential human exposures. This effort involved directed Internet searches and telephone interviews of knowledgeable experts in academia, industry, and Government. The list of documents from the Phase 1 effort was expanded by the Phase 2 results.

The quality assurance/quality control plan for this study included (1) collaborative review of literature survey results, retrieval strategies, and information sources that were developed from guidelines of the Reference and User Services Association of the American Library Association, (2) external reviewers who reviewed the search strategy for completeness, (3) NAL/NRC meetings to review progress and comment, and (4) archives of all online search activities that will be maintained by NAL for 5 years.

This investigation turned out to be very complex for many reasons. For example, there is a paucity of reliable and documented information. Moreover, this report is a *first of its kind* effort to address the realistic material flow of reused soils.

Although this study revealed that no methodology currently exists for formulating realistic soil reuse exposure scenarios and assigning the appropriate parameter values, it did yield some noteworthy observations:

• Large-volume transactions (such as 1,000 cubic yards or more) are generally distributed to construction projects (e.g., roads, parking lots, building pads).

• Large-scale greenhouse and landscape operations have moved away from using natural soils to specially formulated soilless growth media.

• Market leaders in packaged potting "soils" use mineral and organic matter in their formulations.

• Medium to small greenhouse and landscape operations continue to use natural soil mixtures.

• Small-volume transactions involving natural soils are primarily distributed to small businesses and private home owners.

• There still remains a demand for free or low-cost soil for diverse uses by the general public.

• Many landfills restrict some categories of materials for disposal such as yard wastes. This necessitates alternative uses of these materials.

In addition, the study found the following trends in greenhouse and landscape operations:

• Industry seeks inexpensive and readily accessible alternative materials in place of reused soil.

• Limited and variable fertility associated with natural soils encourages the use of specially formulated soilless growth media.

• The high cost of long-distance transportation and the low inherent economic value of the reused soil encourages use of locally available sources.

This study yielded the following benefits:

• This study supported and complimented the scoping scenario characterizations under evaluation in a parallel NRC effort for dose modeling analyses pertaining to soil reuse.

• This study refined parameter values to reflect realism in modeling exercises, and further improvements are expected.

• This study emphasized the significant uncertainty in material flows and soil reuse characterization (i.e., probabilistic distributions of uncertainty are needed).

This study also identified the following human factor sources of uncertainties:

• Material flow and exposure scenarios are a function of human behavior, which varies because of geographic and environmental aspects of the locations where the soil reuse occurs.

• Anecdotal sources were useful because of the shortage of documented information.

This report compiles, for the first time, data and information sources for parameters specific to soil reuse. This report also provides information that is relevant for generic, as well as site-specific dose assessments. This investigation was coordinated with a companion effort by the Federal Interagency Steering Committee on Radiation Standards (ISCORS) study directed toward the disposition of sewage sludge.

Appendices to this report provide more detailed information: Appendix A discusses the public comments received on Draft NUREG-1725; Appendix B presents the Phase 1 detailed search strategies and results; Appendix C presents the Phase 2 additional resources; Appendix D lists general information on soils; and Appendix E identifies the NAL investigators.

FOREWORD

This technical report, NUREG-1725, was prepared by the National Agricultural Library (NAL) staff and the U.S. Nuclear Regulatory Commission (NRC) staff in the Office of Nuclear Regulatory Research and the Office of Nuclear Material Safety and Safeguards. The NAL staff performed the research work under an interagency agreement (RES-99-005 JCN Y6227) with the NRC's Office of Nuclear Regulatory Research. The report provides information on the results of a comprehensive information search related to soil reuse. The NAL research study was undertaken to support the NRC staff in its development of technical bases for defining soil reuse scenarios. The NRC staff considered public comments received on Draft NUREG-1725 in preparing this final report.

It should be noted that this report is not a substitute for NRC regulations, and compliance is not required. The information search strategies and findings documented in this NUREG-series publication are provided for information only. The U.S. Government does not warrant or assume any legal liability or responsibility for the accuracy, completeness, or usefulness of any information, apparatus, product, or process disclosed. Use of product or trade names is for identification purposes only and does not constitute endorsement by the NRC or NAL. In addition, the Uniform Resource Locators (URLs) used to identify Web sites referenced in this report were valid as of the indicated date, or the publication date of this report.

Farouk Eltawila, Director
Division of System Analysis and Regulatory Effectiveness
Office of Nuclear Regulatory Research

ACKNOWLEDGMENTS

The NAL and NRC staff responsible for compiling this report, gratefully acknowledge Mr. Thomas E. Smith, Team Leader, NRC Technical Library, and his staff (particularly Mr. Charles Gorday) for their assistance in obtaining many of the publications listed in this report. Many of these reports and articles were obtained through interagency loans, and we acknowledge the excellent cooperation between the NRC Technical Library staff and their colleagues in the libraries of other Federal agencies.

We also acknowledge the NAL information researchers identified in Appendix E who conducted the information searches, and the external peer-reviewers chosen by NAL: Ms. Carla Long Casler, University of Arizona; Dr. Rufus Chaney, USDA/ARS; Mr. Eric Koglin, U.S. Environmental Protection Agency; Mr. Robert Lacey, U.S. Army Corps of Engineers; Dr. Dewayne Mays, USDA/NRCS; and Ms. Carol Reese, American Society of Civil Engineers.

In addition, we acknowledge the NRC publication review staff (particularly Ms. Paula Garrity, a Technical Editor in the NRC Office of the Chief Information Officer), for their helpful reviews of the manuscripts and suggestions on preparing this report.

Finally, we acknowledge the diligent efforts of Mr. Jon Peckenpaugh, NRC/NMSS, in Phase 1 of this study.

ABBREVIATIONS

AASHTO	American Association of State Highway and Transportation Officials
AEC	Atomic Energy Commission
ARS	Agricultural Research Service (USDA)
ASCE	American Society of Civil Engineers
ASM	available soil moisture
ASTM	American Society for Testing and Materials
ATTRA	Appropriate Technology Transfer for Rural Areas
BLM	Bureau of Land Management (DOI)
BLS	Bureau of Labor Statistics
CAB	CAB International (formerly known as Commonwealth Agricultural Bureau)
CC	category codes (a Dialog® operator)
CD	compact disc
CD-ROM	compact disc, read-only memory
CFR	*Code of Federal Regulations*
CIS	Congressional Information Systems
CISTI	Canada Institute for Scientific and Technical Information
DE	descriptor (a Dialog® operator)
DOC	U.S. Department of Commerce
DOE	U.S. Department of Energy
DOI	U.S. Department of the Interior
DTIC	Defense Technical Information Center
EPA	U.S. Environmental Protection Agency
ERDA	Energy Research and Development Administration
ERS	Economic Research Service (USDA)
FRN	*Federal Register* notice
HAC	Housing Assistance Council
IAA	Interagency Agreement
IAALD	International Association of Agricultural Librarians and Documentalists
IAEA	International Atomic Energy Agency
ID	identifies (a Dialog® operator)
ISCORS	Federal Interagency Steering Committee on Radiation Standards
MSC	Mulch and Soil Council (formerly known as the National Bark and Soil Producers Association)
NAL	National Agricultural Library (USDA)
NASS	National Agricultural Statistics Service (USDA)
NAWS	National Agricultural Workers Survey

ABBREVIATIONS (Continued)

NBSPA National Bark and Soil Producers Association (now known as the MSC)
NCRP National Council on Radiation Protection and Measurements
NEA Nuclear Energy Agency (OECD)
NIOSH National Institute for Occupational Safety and Health
NMSS Office of Nuclear Materials Safety and Safeguards (NRC)
NRC U.S. Nuclear Regulatory Commission
NRCS Natural Resources Conservation Service (USDA)
NTIS National Technical Information Service (DOC)

OCLC Online Computer Library Center, Inc.
OECD Organization for Economic and Cooperative Development
OSHA Occupational Safety & Health Administration

QA/QC Quality Assurance/Quality Control

RD Respirable Dust
RES Office of Nuclear Regulatory Research (NRC)
RUSA Reference and User Services Association, American Library Association

SH subject headings (a Dialog® operator)
SM Statistical Masterfile
STIC Scientific and Technical Information Center
STN STN International

TI title (a Dialog® operator)
TNLA Texas Nursery & Landscape Association

UNSCEAR United Nations Scientific Committee on the Effects of Atomic Radiation
URL Uniform Resource Locator
USAIN U.S. Agricultural Information Network
USDA U.S. Department of Agriculture
USGS U.S. Geological Survey

(F) field (a Dialog® proximity operator)
(N) near (a Dialog® proximity operator)
(S) subfield (a Dialog® proximity operator)
(W) with (a Dialog® proximity operator)

1. INTRODUCTION

1.1 Background

On June 30, 1999, the U.S. Nuclear Regulatory Commission (NRC) published, for public comment, an issues paper indicating that the agency was examining its approach for control of solid material. Specifically, the issues paper presented alternative courses of action for control of solid materials that have very low amounts of, or no, radioactivity.

In August 2000, following public comment on the issues paper, the NRC decided to defer its final decision on whether to proceed with rulemaking on control of solid materials while it requested a study by the National Academies of possible alternatives for control of slightly contaminated materials. During the time that the National Academy of Sciences study is ongoing, the Commission directed its staff to continue developing the technical information base necessary to support a Commission policy decision in this area.

As part of this decisionmaking, it is useful to have information for estimating the potential radiological exposure that could occur if solid material is removed from NRC-licensed facilities. This report focuses on a particular solid material — soil[1].

As part of this ongoing effort, the NRC published for public comment draft NUREG-1725, "Human Interaction with Reused Soil: A Literature Search," in June 2000. A *Federal Register* notice (FRN) published on July 19, 2000, announced the availability of this draft and requested public comments.

---

[1]As used in this report, "reused soil" means soil at or originating from an NRC-licensed site (or Agreement State-licensed site) that may contain small residual levels of radioactivity.

1.2 Scope

This NUREG-series publication reports the results of a 2-year information search focused on human interactions with reused soils in the United Sates. This investigation sought to compile information intended to support the formulation and characterization of scenarios related to exposure to residual radioactivity in reused soils. Developed under an interagency agreement between the NRC and the National Agricultural Library (NAL) of the U.S. Department of Agriculture's (USDA) Agricultural Research Service (ARS), this report documents the process and results of an extensive literature review and focused information search on human contact with soil. This report was issued in June 2000 as Draft NUREG-1725, "Human Interaction with Reused Soil: A Literature Search," to solicit public comments and additional information.

This report provides information on the ways in which soils are reused in commerce (e.g., landscaping) or by the general public in the United States. This information is an important part of the technical basis for assessing the possible exposures that could result if soil is released from NRC-licensed facilities. Specifically, this information will be used to develop soil reuse scenarios for use in dose assessments. To develop these scenarios, it is necessary to obtain data on the forms of human contact with soils in the United States. Using this information, the NRC staff and contractors are working to define realistic soil reuse scenarios and to estimate parameters for simulating exposure pathways involving soil removed from NRC-licensed facilities.

NRC staff and researchers from the NAL conducted this investigation in two phases.

For the initial literature search in Phase 1, the information sources searched included the Dialog® (an online system of more than 500 databases), the Internet, and other sources. With support from NRC staff and external reviewers (non-NAL library and information science and soil science professionals), the NAL researchers developed targeted search strategies to retrieve relevant items from the Dialog® databases. These search strategies were structured to query literature from science publishers, academic presses, professional societies, trade journals and bulletins, theses and dissertations, as well as information published in industry standards, newspapers, company reports, statistical sources, etc. The Dialog® search was developed by creating three main categories or concept sets, including "General" (actions or activities of humans with soils), "Particular" (specific identified types of human-soil interactions), and "Volume" (volume, quantity, or economic terms that quantify or delineate the extent of human contact with soil).

In the Phase 1 effort, the researchers screened the initial records found in the Dialog® database searches for relevance to the scope of this research effort. Through the detailed review of these records by NAL and NRC researchers, a total of 56 documents were identified as relevant.

In the Phase 2 effort, the emphasis was on identifying information sources for formulating and estimating parameters to be used in modeling soil reuse scenarios involving potential human exposures. This effort involved directed Internet searches and telephone interviews of knowledgeable experts in academia, industry, and Government. The list of documents from the Phase 1 effort was expanded by the Phase 2 results.

This report supersedes Draft NUREG-1725 in identifying information on ways in which soils are reused in commerce or by the general public. This information will then be used in assessing the possible exposures that could result if soil is removed from NRC-licensed facilities. This

information will assist in developing a reasonably complete characterization of relevant usages for these reused soils.

These soil reuse scenarios will include, but not be limited to, soil processing, construction, agriculture, and various commercial and residential uses of reused soil and soil-related products. A parallel effort by a Federal interagency working group is examining exposure scenarios related to use of sewage sludge and ash. Although the sewage sludge project is outside the scope of this report, the soil reuse effort has been harmonized with it.

NUREG-1725 is intended as input into the NRC's decisionmaking process, with respect to the potential environmental impacts from reuse of such soils. It also serves as a basis for identifying site-specific scenarios for the purpose of dose modeling. Although soil radioactivity was not within the scope of the search, it is being addressed in the NRC staff's analysis of potential exposure scenarios from soil reuse.

This report does not address acceptable disposition of such soil materials, nor does it endorse any particular commercial use of such material.

1.3 Public Comments

In response to two FRNs issued on July 19 and September 18, 2000, the NRC received 190 public comments. A discussion of these comments appears in Appendix A. Although many of the comments were outside the scope of the draft report, they will be considered in the NRC's future deliberations.

1.4 Purpose of this Report

The purpose of this report is to identify the technical basis and associated information sources for characterizing the ways in which soils are reused in commerce (e.g., landscaping) or by the general public. This information can then be used in assessing the potential radiation exposures that could result if soil is removed from NRC-licensed facilities.

As part of this technical basis, it was essential to obtain available information on the reuse of excavated soils in the United States. Using this information, the NRC staff will conduct exposure pathway modeling, which will reflect a realistic range of potential scenarios consistent with realistic reuses of soil.

This report describes the methodologies used in developing the information search strategies and in selecting the database sources. These strategies and sources were reviewed by external reviewers, whose comments are included in the report. The primary focus of the work was to identify documented, verifiable references for the NRC staff. Therefore, the principal focus of this study was to search the published literature.

The information sources searched for this report include the collections of the NAL; Dialog®, an online system of more than 500 databases; the Internet; and other sources. On the basis of the surveyed literature, the NAL staff made recommendations to the NRC staff, who selected a subset of the documents for further analysis. These documents are listed in Table 4.2.

1.5 Content of this Report

Section 1 presents the motivation for the research study, background information, scope and intended purpose. Section 2 identifies the information categories sought for characterizing soil reuse and environmental pathways. Section 3 details the information search process for the two research phases. Section 4 presents the Phase 1 initial literature search strategy, literature and Internet sources identified and results. Section 5 details the focused information search identifying U.S. and international information sources, and Web sites. Section 6 outlines the quality assurance and quality control plan. Section 7 summarizes the significant findings of the 2-year research. Section 8 provides a listing of references that were cited throughout the report. Appendices to this report provide more detailed information: Appendix A discusses the public comments received on Draft NUREG-1725; Appendix B presents the Phase 1 detailed search strategies and results; Appendix C presents the Phase 2 additional resources; Appendix D lists general information on soils; and Appendix E identifies the NAL investigators.

2. INFORMATION CATEGORIES CHARACTERIZING SOIL REUSE SCENARIOS AND ENVIRONMENTAL PATHWAYS

2.1 Introduction

This research is intended to yield information for use in characterizing the ways in which soils are reused in commerce or by the general public. This information can then be used in assessing the possible exposures that could result if NRC licensees seek to release soil with very small amounts of, or no, radioactivity during routine operations. For example, NRC licensees might seek to release excess soil during construction and regrading activities. In order to assess potential exposures, analysts will need the information cited in this report in order to characterize soil reuse scenarios and environmental pathways.

There is considerable complexity in defining information needs for characterizing and assessing soil reuse scenarios and environmental pathways because of the broad diversity of NRC and Agreement State licensees, and the nature of the case-by-case approach to evaluate and approve the removal of soils from these diverse regulated facilities.

2.2 Soil Reuse Scenarios

Soil reuse scenarios include, but are not limited to, agriculture, landscaping, soil processing, construction, and recreational activities. These scenarios were identified by the NRC staff following review of the Phase 1 literature search results. The NRC staff anticipated that these scenarios reflect most of the human interactions with reused soils. Depending upon the specific exposure scenario to be modeled, relevant information may include such aspects as human contact time, soil composition, number of people interacting with reused soil, and volumes of soils (see Table 2.1).

Table 2.1 Typical Parameters Used to Characterize Soil Reuse Scenarios
Number of people interacting with soil
Proximity of people to reused soil
Range of hours per day spent in close proximity to soil
Range of days per year in contact with reused soil
Volume of soil involved in the activity
Breathing rate per hour
Daily water consumption
Indigenous fruits and vegetables consumed
Range of dust sizes and amount inhaled
Site-specific climatic and geographical characteristics (e.g., growing season)
Soil composition (e.g., % of soil additives)
Note: Although soil radioactivity was not within the scope of the search, it is being addressed in the NRC staff's analysis of potential exposure scenarios for soil reuse.

Information cited in this report can be used to model soil reuse scenarios (such as construction backfilling around a residence), and activities of the people living in the residence (such as landscaping), could be used to define a "rural resident scenario." Additional aspects of scenario formulation include the detailed interactions of humans with the reused soil to define exposure pathways.

Table 2.2 identifies a range of possible human activities involving reused soils. The NAL and NRC staff investigators used these categories to organize their information search strategies on how soil may be reused and to collect specific details on how people may come into contact with it. This table was useful in guiding the information search and organizing the parameter studies (see Section 5).

During the information search, as more information became available on current reuses of soil, parameters such as those listed in Tables 2.1 and 2.2 were being sought for the scenarios. Section 5 presents detailed discussions concerning these parameters and the related search findings. These tables proved to be useful dynamic tools, which evolved during the iterative search process.

Table 2.2 Possible Human Activities Involving Reused Soils
Greenhouse production
Farm and field agriculture
Landscaping
Soil processing
Construction
Transportation
Landfills
Recreational [1]
[1] Note: This scenario was not researched in detail, since the Phase 1 literature indicated that the landscape workers preparing and maintaining the recreational venues would be the most likely individuals to spend the most time and be in routine close proximity to the reused soil.

2.3 Environmental Pathways

In order to analyze the exposure scenarios, it is necessary to identify the pathways by which humans come into contact with the reused soil. For example, for gardening activities within the suburban scenario, exposure pathways would include inhalation, ingestion of vegetables or fruits, inadvertent ingestion of soil, and external exposure.

The complete scenario may have more than one set of exposure pathways, depending on the expected activities of the residents. For example, the resident may have a home office or bedroom in the basement, with the associated external exposure pathway from soil used as backfill. Therefore, the search included information on all of the applicable pathways, including intake quantities and exposure times.

Figure 1, adapted from the user's manual for RESRAD Version 6 (Yu et al., 2001) portrays these potential environmental pathways for human exposure. NUREG-1725 provides information sources (generic and specific) for formulating and estimating parameters for modeling soil reuse exposures using a multimedia environmental dose assessment model such as RESRAD.

**Figure 1.
Exposure
Pathways
Considere
d in
RESRAD
(Yu et al.,
2001)**

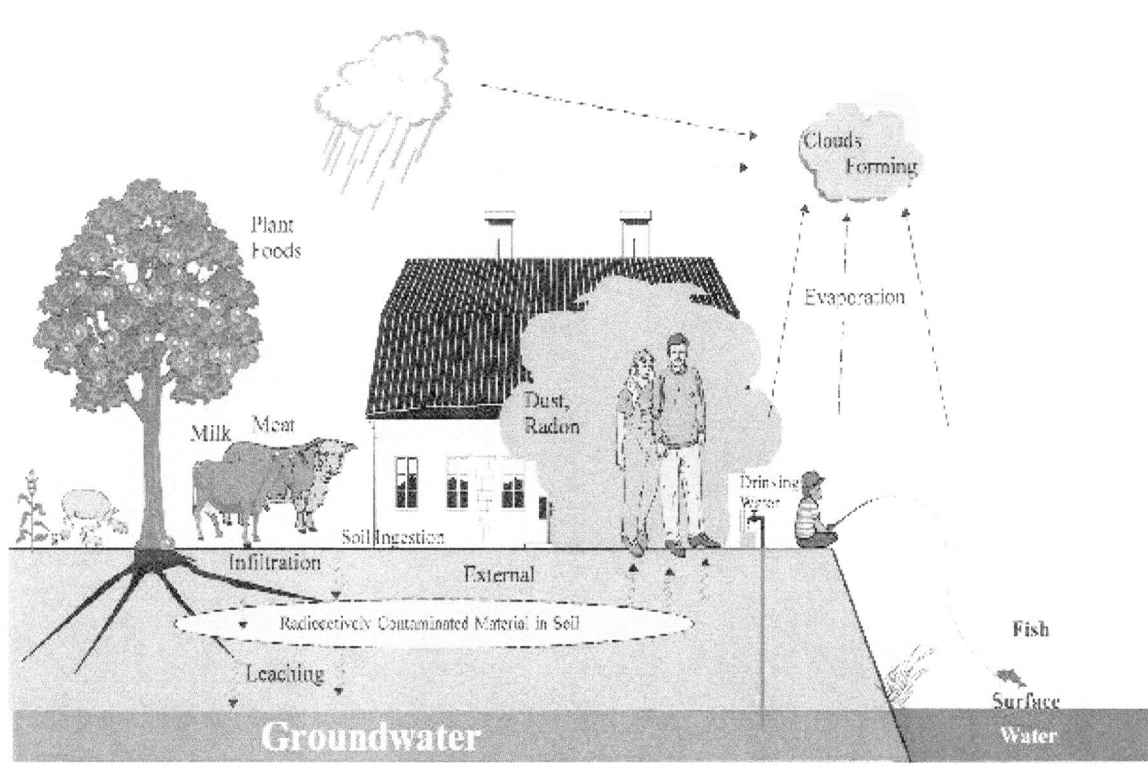

3. INFORMATION RESEARCH PROCESS

3.1 Introduction

This study was conducted in two distinct phases. The first phase was a broad sweep of the published literature to review the extent and nature of information sources documenting human interactions with reused soil. The second phase of the study focused on identifying information resources for the specific parameters listed in Tables 2.1 and 2.2.

Bibliographic databases were the primary information resource searched in Phase 1. The structured nature of the databases allowed for development of a highly structured research approach. Both the approach and results were thoroughly reviewed. Phase 1 results indicated a paucity of systematic studies on human interactions with reused soil in the published literature.

As a result of the Phase 1 findings, the research approach for Phase 2 was modified. Specifically, Phase 2 searched broader information resources, relying to a lesser extent on bibliographic databases and to a greater extent on the Internet and direct contact with experts and industry representatives. These contacts were often used as an integral part of the quality assurance (QA)/quality control (QC) for the Phase 2 study, which by its nature was not formally structured.

3.2 Interagency Agreement

The research project began in July 1999 with exploratory meetings held at NAL. The research scope, mutuality of interest, and staff expertise were established by all parties at these meetings. The understanding that was reached was formalized through an interagency agreement (IAA) dated August 20, 1999, which spelled out the scope of the project in detail.

As stated in the IAA, the objectives were to (1) conduct focused information searches to support the formulation and characterization of scenarios related to reuse of soils removed from NRC-licensed facilities, (2) develop a draft NUREG report for public comment on the literature findings, and (3) produce a final NUREG report documenting the search results. In summary, the information sought in this study was a reasonably complete characterization of relevant soil reuse. The NRC staff will use this information to characterize exposures to soils that may be excavated and transported from NRC-licensed facilities for use in commerce or by the general public.

3.3 Phase 1 Methodology

Information research is conducted in iterative stages beginning with an initial problem statement. In the next stage of the process, information professionals create search strategies and select databases and other information resources. Next, the strategies are run against the selected databases and the initial results are analyzed for relevancy and completeness. The initial results are then reviewed with the client to elicit feedback. The process then repeats, beginning with refinements to the search strategy and database selection.

For example, the NRC staff was interested in quantitative information on temporary and long-term storage of soil. A search strategy was constructed to first gather records that contained one or more of the terms: soil, soils, or dirt (soil or soils or dirt). Next a search was constructed to gather records containing one or more of the terms: storage, storing, dispose, disposed, disposal (storage or store or dispos???). Then the two sets of records are combined and filtered using the Boolean operator "and" — [(soil or soils or dirt) and (storage or store or dispos???)]. Only those records containing terms from both sets pass through the filter. This collection of filtered records is called a concept set. Please refer to Appendix B for an in depth discussion on information retrieval methods and strategies used in this study.

On the basis of discussions between the NRC and NAL staffs, sets of search strategies were initially created and organized into three broad conceptual categories, including general concepts, particular scenarios, and volumetric studies. The early strategies were reviewed with the NRC staff. Additional terms were added to the strategies, and the concept sets were ranked and prioritized at that time (see Table 4.1).

Over the next 6 weeks, the strategies were run against selected databases, and record titles were downloaded and shared with the NRC staff for their review and selection. The NRC staff actively participated through telephone, electronic mail, and face-to-face meetings.

The literature search strategies were structured to query literature from science publishers, academic presses, professional societies, trade journals and bulletins, theses and dissertations, as well as information published in industry standards, newspapers, company reports, and statistical sources.

3.3.1 Phase 1 Internal Review   Process

A team of information professionals was assembled to work collaboratively on the NRC project (see Appendix E). The team reviewed the primary search strategies with the NAL principal investigator responsible for comprehensive search retrieval. Based on team comments, search strategies were refined prior to final execution.

The comprehensive searches conducted in the Phase 1 study were designed to retrieve all potentially relevant literature. This process retrieved a large number of titles for consideration.

The large number of items retrieved through the comprehensive search is partly attributable to the

inclusion, within the search strategy, of nonspecific terms (such as "use") that occur frequently within bibliographic records. To facilitate NRC staff and external review, the team used the following three modifications of the primary search methodology to further refine and reduce the total number of titles retrieved and to improve overall relevancy:

(1)   The strategies were made more selective for example by using field limitations (see Appendix B, Section B.2.21).

(2)   Databases selected for searching were focused on 10 to15 highly productive files.

(3)   NAL staff reviewed the primary search results and pre-selected titles for the draft letter report.

Selected results in the Phase 1 draft letter report, along with files containing the complete comprehensive search findings were provided to the NRC staff. A second round of selections were made by the NRC staff for the Phase 1 final report.

3.3.2 Phase 1 Reports and Products

NRC staff selected candidate titles for further consideration. Complete bibliographic citation[2] information were retrieved for these candidate titles. When available the citation information included abstracts. The additional information enabled the NRC staff to perform a second round of selections for Draft NUREG-1725.

Bibliographic citations include information about book and journal titles, volumes, page numbers, and publication dates. Using this information, copies of full-text reports can be obtained by request from libraries, bookstores, Government

---

[2] Use of the term citation throughout this report refers solely to information describing publication authorship and source. The term is not used in any legal or regulatory sense.

agencies, publishers, or other access providers.

Database providers typically charge customers for complete bibliographic citation access. For the purposes of this project, it was not cost-effective in the Phase 1 comprehensive search. It was mutually agreed that NAL would provide in the Phase 1 final letter report dated November 1999, complete citation information for each of the NRC staff selections.

The draft letter report, dated November 1999, included the following information:

• instructions to and list of external reviewers

• comprehensive search strategy concept sets

• bibliography of selected titles

Although the primary focus of the IAA between NAL and the NRC was to locate citable information from peer-reviewed published literature, it was also agreed that NAL would search the Internet. Preliminary results were cited in the Phase 1 draft letter report. A more thorough search of the Internet was conducted in the Fall of 1999. The results of these searches were given to the NRC staff in a notebook identified as a "Supplement to the Draft Letter Report."

A final letter report, concluding the work for Phase 1, was provided to the NRC staff in November 1999. This final letter report was revised and then published as Draft NUREG-1725 in June 2000.

3.4 Phase 2 Methodology

The NRC team reviewed the Phase 1 research findings and evaluated the information needs for a dose assessment analysis. The NRC staff then ranked those Information needs for anticipated exposure to reused soil and priority of information needed. Agricultural and landscape workers were ranked high in both categories.

In March 2001, a larger team of NAL information research specialists was called upon to study the specific parameters identified by the NRC staff. Meetings were held with NAL, NRC staff, and contractors at timely intervals. At these meetings, progress was reviewed and strategies were refined.

Specific goals for the Phase 2 study were: (1) to identify citable references and sources with quantitative data related to the reuse of soil; (2) to query scientists, extension specialists, and business representatives about actual practices; and to provide examples of typical data that might be used in future modeling.

A broader research approach was necessary for the Phase 2 objectives. Phase 2 researchers conducted initial literature searches for their assigned parameters using techniques similar to those described in the Phase 1 study. The Phase 2 literature searches identified information sources, experts, and empirical observations.

In addition to conducting literature searches, NAL researchers consulted Government Web sites for statistical information and relevant standards and regulations. Professional and trade organization Web sites were reviewed as well. These searches yielded many valuable resources. In the absence of published studies, indirect sources were sought. These indirect sources were studies where the main objective was not soil reuse but could provided related information for estimating parameter values. This information was found in published papers, through extensive Internet searches, and by telephone and email interviews with academic and industry experts.

Typical data were extracted from these diverse information sources and are presented in Section 5. Wherever possible two or more sources were compared. Finally, scientists, extension agents, industry representatives and other professionals were consulted as a final validation for the reported trends and data.

The NRC staff contributed additional information resources through independently conducted Internet searches, professional contacts, and a review of national and international regulatory statutes and standards. The NRC staff concentrated on parameters relating to civil engineering, and transportation. NRC staff results are included in Section 5.6 of this report.

3.4.1 Phase 2 Review Process

Two factors that significantly impacted the Phase 2 review process were (1) the diverse and unstructured nature of the information resources on the Internet, and (2) the lack of directly relevant published studies. Therefore, the Phase 2 internal review processes relied upon independent validation between anecdotal references and consultation with experts as described above.

In Phase 1 the external peer-review process concentrated on a review of the research process. In Phase 2, the external peer-review dealt primarily with the technical content and interpretation of the overall study. Section 6 describes the external peer review for this study.

3.4.2 Phase 2 Reports and Products

A briefing for NRC staff was held at the NAL on August 30, 2001. At the briefing, the Phase 2 draft letter report documenting the major findings of the Phase 2 study was presented. NAL and NRC staff working together used the letter report which was finalized on October 5, 2001 to produce this final version of NUREG-1725. The report summarizes 2 years of research to establish a knowledge base of information sources about human interactions with reused soil. It brings together the first-of-its-kind research findings from both NAL and NRC information research processes.

4. PHASE 1: INITIAL LITERATURE SEARCH

4.1 Strategy Development

In July 1999, the NRC staff contacted NAL to explore the possibility of having NAL staff assist the NRC in developing comprehensive search strategies to locate and document citable sources covering every aspect of soil use that might be applicable to the release of soils from NRC-licensed sites.

Specifically, the NRC staff asked the NAL staff to document the scope of this broad topic area by conducting a general survey of the NAL collection through its AGRICOLA database and other databases available from Dialog®, a system of more than 500 databases covering science, technology, business, news, and other categories. This initial survey yielded well over two million records that contained the terms "soil" or "soils", or "dirt" or "earthen" in the title or as subject keywords. Further examination of the records indicated that the vast majority of the records were not relevant for the NRC project. After detailed discussions with the NRC staff, NAL developed targeted search strategies to retrieve relevant items.

NAL also established general parameters to define the scope of the searches. For example, all literature published on relevant topics going back as far as 1970 was sought. English language literature was specified, although non-U.S. publishers were not excluded. In addition, the parameters excluded items covering normal soil testing operations for plant nutrients, pH, cation exchange capacity (CEC), bulk density, etc.

To stay within NRC-specified parameters, NAL defined two additional exclusions. First, because the soils would remain in the United States, export data on potting soil, for example, was not sought. Second, because the focus involved the use or treatment of soil removed from a native site, studies treating soils in normal farming and agricultural settings were not searched. Further,

because even these targeted searches retrieved thousands of items, it was agreed that the initial report of search results would be limited to record titles only. This abbreviated title format allowed the NRC staff to scan records and select specific items for complete citation access and further review.

Initial work on strategy concept development and term selection was based on concepts and terms presented by the NRC in the statement of work for this project. These basic concepts and terms were extended on the basis of NAL staff expertise with soil science topics and operations and with natural resource literature. The search strategies developed by the NAL principal investigator were internally reviewed and then shared with the NRC staff. The preliminary terms and delimiters were incorporated in the search strategy and concept set constructions. This established priority rankings before complete searches and title downloads were executed. The final search strategies used are shown in Appendix B.

4.2 Database Selection Process

Database selection began using the DIALINDEX® file from Dialog® to identify databases that contained records with specified search terms and showed the numbers of records that would be retrieved by those terms in each database. The databases available from Dialog® may be grouped in the DIALINDEX® system into subject and source categories, including "allscience" (258 files), "allbusiness" (348 files), and "allnews" (160 files). Relevant databases were initially selected for search by reviewing the number of items posted for each concept set. These major subject categories together comprise more than 470 unique database files[3] (many files are included in more than one of the aforementioned categories).

---

[3] DIALINDEX® includes most but not all of the files available through the Dialog Corporation.

Counts of items retrieved from each database for all the defined concept sets were the initial results used in database selection.

Databases with significant item counts were further analyzed for sources of records, scope of subject coverage, and time span of records included. The general project parameters were used in the database selection process. The following databases were excluded:

- business, science, and news databases focused outside North America

- business databases focused on personnel and management topics

- business databases focused on stocks, corporate finance and mergers

- newspaper databases from locations other than North America

- newspaper databases[4] added to Dialog® after 1990

- science databases for non-subject topics (biotechnology, computer science and programming, mathematics).

From the more than 470 unique databases in the major groups surveyed for file inclusion, 200 databases were specifically selected and actually utilized in this study. The database file names with their associated Dialog® file numbers, dates of record inclusion, and updates are shown in Appendix B.

For specific concept sets (Section 4.3 and Appendix B), major category groups of database files were sometimes excluded. For example, most business files were not searched for soil material flow or soil remediation/reclamation methods, and

---

[4] Recently added newspaper files were generally local papers. Major stories carried in local papers are also covered through national wire services that were already included among the databases searched.

economics terms were not searched in all science databases.

Other file selection decisions were made after initial search efforts, including the decision that patent databases would not be included. NAL and NRC staff agreed that patents involving soil with impact in terms of soil-human interactions would likely be documented in other business, science, or news database files. This was shown to be the case in several specific instances, with patents retrieved from these other sources.

Although database files focused specifically on geographic areas outside the United States or Canada were generally excluded, much literature from non-U.S. examples of soil-human interactions was retrieved. Items selected by the NRC staff included citations treating soil topics involving Chernobyl and other non-U.S. activity sites and non-U.S. publishing locations.

Index databases for newspapers covering multiple titles and wire service index files were also searched for many topics, as well as popular literature index databases such as Magazine Index® and Reader's Guide®, covering generally nontechnical sources. Specific databases included for the searches for each concept set are shown in Appendix B.

The concept sets developed to search the literature are organized into three strategy groups, each of which contains three or more specific concept sets, as listed in the next section. Appendix B presents the detailed strategy statements, with notes that describe the concept set purposes and the terms, codes, and Dialog® commands that appear in the search statements and database selections. Proximity operators that are shown were sometimes adjusted to increase stringency during final retrieval operations from the initially approved strategy sets to improve the overall relevancy of the results.

Dialog® databases are structured information resources. Each database has specific fields such as descriptors (DE), identifiers (ID), abstracts (AB), and others included in the basic index for each file. Searches retrieve records in which a term is found in any basic index field unless particular fields are listed in the search command to limit retrieval. When it is desired that a term appears only in one or more specific fields, the syntax "term / f1, f2, f3", etc., may be used, with selected term tags (DE, ID, etc.) entered after "term/" , in place of "f1, f2, f3".

Dialog® allows other sophisticated search and retrieval commands. Those called proximity commands are based on relative locations between terms or parenthetically nested groups of terms. Standard Boolean logic commands may be utilized to further specify retrieved item characteristics. Boolean set [5] combination commands include "and," "or," and "not." These operators link terms or groups of terms in a search statement by requiring their respective occurrence to either "must be", "may be," or "must not be," included in records retrieved.

4.3 Literature Survey

The literature survey was organized into three broad categories for searching, including general (labeled G), particular (labeled P), and volumetric or quantitative (labeled V). Detailed descriptions of the actual search strategies, notes on the Dialog® command syntax, and databases selected for each set are presented in Appendix B. Refer to Appendix B for a complete listing of the files searched for their file names and coverage dates.

4.3.1 Search Category G (General)

Search category G was designed to discover any activities reporting how humans use soil. The terms used in these sets were broad and nonspecific. The group contained six sets that examined concepts including commercial material

---

[5] A Boolean set refers to terms or groups of terms that are connected using Boolean operators (i.e., "and," "not," "or").

flow, storage, processing, general use, and Government publications. Altogether, the six concept sets retrieved a total of 21,310 items for NRC staff review.

4.3.2 Search Category P (Particular)

At the outset of the research project, certain scenarios describing human uses of excavated soils were known. Search category P was designed to retrieve records relating to these known scenarios. Scenarios in the category relate to construction uses; pottery; recreational uses; dust; potting, garden, and topsoil; and so forth. Search category P comprised 11 concept sets and retrieved 27,296 records for NRC staff review.

4.3.3 Search Category V (Volume)

Parameter information needed for dose modeling studies includes contact time, number of people involved, volumes of soil, and so forth. Search category V was designed to discover relevant parameter information. The sets searched for information relating to soil as a commodity, transportation of soils, and statistical information. Search category V comprised three concepts sets and retrieved a total of 29,271 records for NRC staff review.

4.4 Extent of Available Literature

From more than two million database records initially found in surveys of Dialog® databases, approximately 78,000 titles were provided to the NRC staff for review. Table 4.1 summarizes the number of titles retrieved for each concept set. Additional information on search methodology and search results is provided in Appendix B.

After NRC staff reviewed titles from the exhaustive literature search results, selections were made for obtaining complete citation information. Using this information, the NRC staff selected 269 specific items for further review and detailed study. NAL compiled a complete bibliographic listing of these selected items for the Phase 1 final letter report.

Table 4.2 includes the final selections by the NRC staff, plus additional citations identified following the Phase 1 literature search which were reported in the Draft NUREG-1725.

Table 4.1 Summary of Concept Set Findings

Concept Set [a]	Total Bytes	Total Records
G1 *Soil use*	2,992,239	12,424
G2 *Soil material flow*	116,521	502
G3 *Soil process (not soil forming)*	146,378	719
G4 *Human contact with soil*	825,283	2,404
G5 *Storing soil*	1,082,646	4,966
G6 *Soil Publications from applicable Federal agencies*	64,237	295
P1 *Golf courses and sods*	30,172	150
P2 *Reclamation methods*	1,014,986	5,143
P3 *Soil dust*	118,490	516
P4 *Earthmoving and soil use in construction fill and rammed earth*	697,177	3,388
P5-6 *Soil in walls, dams, berms and dikes*	3,124,201	14,199
P7 *Adobe*	34,755	177
P8 *Pottery production and potting clay*	27,035	152
P10 *Soil erosion rates* [b]	42,236	39
P11 *Potting soil and bagged or bulk soil*	669,537	3,254
P12 *Topsoil*	134,414	278
V1 *Soil economics, business activities*	2,295,349	9,363
V2 *Statistical and numeric data for soils*	654,653	2,745
V3 *Soil transportation*	837,839	17,163
Total *19 concept set results*	14,908,148	77,877

[a] No relevant material was found in concept set P9 (detrital materials).
[b] Related materials from the Natural Resources Conservation Service (NRCS/USDA) Web pages and links were also provided.

Table 4.2 Phase 1 Final Literature Selections

100 Area Hanford soil washing treatability tests.
Westinghouse Hanford Co., Richland, WA, *Department of Energy environmental remediation conference, Augusta, GA (United States), 24–28 Oct 1993,*Department of Energy, Washington, DC, 10 p., September 1993.

100 Area soil washing: Bench scale tests on 116-F-4 pluto crib soil.
Westinghouse Hanford Co., Richland, WA. Department of Energy, Washington, DC. 95 p., June 10, 1994.

^{137}Cs mobility in soils and its long-term effect on the external radiation exposure.
Bunzl -K, Jacob -P, Schimmack -W, Alexakhin -RM, Arkhipov -NP, Ivanov -Y, Kruglov -SV, *Radiation and Environmental Biophysics, 36(1): 31-7,* February 1997.

Absorption of radiocesium by sheep after ingestion of contaminated soils.
Cooke -AI, Weekes -TEC, Green -N, Wilkins -BT, Rimmer -DL, Beresford -NA, Fenwick -JD, *Science of the Total Environment, 192(1): 21–29,* October 8, 1996.

Alternatives for management of wastes generated by the formerly utilized sites remedial action program and supplement.
Gilbert -TL, Peterson -JM, Vocke -RW, Alexander -JK, Argonne National Lab., IL (USA) Department of Energy, Oak Ridge Operations Office, Oak Ridge, TN, 39 p., March 1983.

Ash: A valuable resource. Volume 4. Ash handling/transportation-roads-engineering fill-marketing. Assessing inhalation exposure from airborne soil contaminants.
Council for Scientific and Industrial Research, Pretoria (South Africa). *Presented at Council for Scientific and Industrial Research Conference Centre, Pretoria, South Africa, February 2–6, 1987.*

Assessing inhalation exposure from airborne soil contaminants.
Shinn -JH, U.S. DOE, Washington, DC, Report No. UCRL-ID-130570, 9 p., April 1, 1998.

Bark and soil producers product index.
Lee -SY, Tamura -T, Larsen - IL, National Bark and Soil Producers Association (NBSPA) Membership Directory, pp. 19-21, NBSPA, Manassas, VA, 1998.

Biological and chemical tests of contaminated soils to determine bioavailability and environmentally acceptable endpoints (EAE).
Montgomery -CR, Menzie -CA, Pauwells -SJ., *Society of Environmental Toxicology and Chemistry (SETAC) 17th Annual meeting—Abstract book, "Partnerships for the environment: Science, education, and policy, November 17-21, 1996, Washington, DC", p 198–199, SETAC, 378p.,* Pensacola, FL, 1996.

Building with adobe brick.
Masterson -R, *Studio Potter, 4 (2): 54–58,* 1975.

Table 4.2 Phase 1 Final Literature Selections (continued)

Calculation of soil cleanup criteria for carcinogenic volatile organic compounds as controlled by the soil-to-indoor air exposure pathway.
Sanders -PF, Stern -AH, Environmental-Toxicology-and-Chemistry, 13(8): 1367–1373, 1994.

Characteristics of radionuclide-contaminated soils from the Sedan crater area at the Nevada test site.
Lee -SY, Tamura -T, Larsen -IL, Essington -EH, *Soil Science, 144(2): 113–121*, August 1987.

Chemical contaminants in house dust: Occurrences and sources.
Battelle, Columbus, OH; Environmetrics, Inc., Seattle, WA; Engineering Plus, Seattle, WA, Funded by Environmental Protection Agency, Research Triangle Park, NC, Atmospheric Research and Exposure Assessment Laboratory, 8 p., 1993.

Clean slate transportation and human health risk assessment.
Department of Energy, Nevada Operations Office, Las Vegas, NV, 60 p., 1997.

Critical pathways of radionuclides to man from agro-ecosystems. Annual progress report October 1980–September 1981.
Smith -MH, Alberts -JJ, Adriano -DC, McLeod -KW, Pinder -JE, III, Savannah River Ecology Laboratory, 50 p., April 1982.

Dermal exposure assessment: Principles and applications. Interim rept.
Versar, Inc., Springfield, VA; Funded by Environmental Protection Agency,
Office of Health and Environmental Assessment, Washington, DC, 392 p., January 1992.

Determination of transfer coefficients for ^{137}Cs and ^{60}Co in a slime-soil-grassland ecosystem.
Handl -J, Kuehn -W, *Health Physics, 38(4): 703–705,* April 1980.

Directory of principal construction sand and gravel producers in the United States in 1997.
U.S. Geological Survey, U.S. Department of the Interior, Mineral Industry Surveys, pp 1–12, Reston, VA, March 1999.

EPA engineering bulletins: current treatment and site remediation technologies.
U. S. Environmental Protection Agency Bulletin, Government Institutes, Inc., 172 p., 1993.

Establishment and maintenance of grassed sports fields - experience from a field experiment on soil construction alternatives. [Original title: Sportgrasytors etablering och skotsel - erfarenheter fran ett markbyggnadsforsok.]
Karlsson - IM, *Rapporter-fran-Jordbearbetningsavdelningen, No. 89,* Department of Soil Sciences, Swedish University of Agricultural Sciences, Uppsala, Sweden, 94 p.,1996.

Federal focus: Army base recycles contaminated soil for pavement.
Mouche -C, *Pollution Engineering, 31(1): 39–40,* January 1999.

Table 4.2 Phase 1 Final Literature Selections (continued)

Field measurement of dermal soil loadings in occupational and recreational activities.
Holmes - KK Jr., Shirai -JH, Richter -KY, Kissel - JC, *Environmental Research, 80(2 Pt 1): 148–57,* February 1999.

First Energy and Barnes Nursery create soils technology, LLC.
Business Wire, Akron, OH, p. 7021143, July 2, 1998.

Fugitive dust emissions from construction haul roads.
Struss -SR, Mikuck i -WJ, Army Construction Engineering Research Laboratory, Champaign, IL, 53 p., 1977.

Geochemistry and mineralogy of soils eaten by humans.
Aufreite -RS, Hancock -RGV, Mahaney -WC, Stambolic -RA, Sanmugadas -K,
International Journal of Food Sciences and Nutrition, 48(5): 293–305, 1977.

Hazardous soil to be used in paving mix.
Civil Engineering News 5(4): 29, Civil Engineering News, Marietta, GA, 1993.

The interactions of low-level, liquid radioactive wastes with soils: 1. Behavior of radionuclides in soil-waste systems.
Fowler -EB, Essington -EH, Polzer -WL, *Soil Science, 132 (1): 2–12,* July 1981.

The interactions of low-level, liquid radioactive wastes with soils: 2. Differences in radionuclide distribution among four surface soils.
Essington -EH, Fowler -EB, Polzer -WL, *Soil Science, 132 (1):* 13–18, July 1981.

The interactions of low-level, liquid radioactive wastes with soils: 3. Interactions of waste radionuclides with soil from horizons of two soil series.
Polzer -WL, Fowler -EB, Essington-EH, *Soil Science 132, (1):* 19–24, July 1981.

Introduction to symposium 19: construction and use of artificial soils.
Koolen -AJ, Rossignol -JP, Kutilek -M (ed.), Horn -R (ed.), Clothier -BE (ed.), State-of- the-art in soil physics and in soil technology of anthrophic soils, Proceedings of the World Congress of Soil Science, Montpellier, France, 20–26 August 1998, *Soil and Tillage Research 47(1-2): 151–155,* 1998.

Issues of risk assessment and its utility in development of soil standards: the 503 methodology an example.
Ryan -JA, Chaney -RL, Issues of Risk Assessment and Its Utility in Development of Soil Standards: The 503 Methodology as an Example, in Prost -R (ed), *Contaminated Soils: Proceedings of International Conference on the Biogeochemistry of Trace Elements, Paris, France, May 15–19, 1995,* pp. 393–413, Colloque No. 85, Institut National de la Recherche Agronomique, 525 p., Versailles, France, 1997.

Large-scale adobe-brick manufacturing in New Mexico.
Smith -EW, *Circular - New Mexico Bureau of Mines and Mineral Resources, (182): 49–56,* 1982.

Table 4.2 Phase 1 Final Literature Selections (continued)

Lead in paint, soil and dust: health risks, exposure studies, control measures, measurement methods, and quality assurance.
Beard -ME, Iske -SDA, (eds), *1993 Boulder Conference on Lead in Paint, Soil and Dust, Boulder, Colorado, July 25–29 1993,* ASTM STP 1226, American Society for Testing and Materials (ASTM), Conshohocken, PA, 422 p., 1995.

Marketing organic soil products.
LaGasse -R, *BioCycle, 33(3): 30–33,* March 1992.

Methodology to estimate the amount and particle size of soil ingested by children: implications for exposure assessment at waste sites.
Calabrese -EJ, Stanek -EJ, Barnes -R, *Regulatory Toxicolology and Pharmacology, 24(3): 264–268,* December 1996.

A Native American exposure scenario.
Harris -SG, Harper -BL, *Risk Analysis, 17(6): 789–95,* December 1997.

National Research Council study targets US soil programs.
National Research Council, *AGROW World Crop Protection News, (198):13,* December 17, 1993.

Off-Site Radiation Exposure Review Project: Phase 2 Soils Program, Revision.
Water Resources Center, University of Nevada at Las Vegas, Las Vegas, NV, Department of Energy Publication DOENV1038423Rev, Funded by U.S. Department of Energy, Washington, DC, 206 p., December 1989.

On the effect of probability distributions of input variables in public health risk assessment.
Hamed -MM, Bedient -PB, *Risk Analysis, 17(1): 97–105,* February 1997.

Probabilistic prediction of exposures to arsenic contaminated residential soil.
Lee -RC, Kissel -JC, *Environmental Geochemistry and Health, 17(4): 159–168,* 1995.

Radiation exposure from radionuclides in ground water: An uncertainty analysis for selected exposure scenarios.
Prohl -G, Muller -H, *Radiation and Environmental Biophysics, 35(3): 205–218,* August 1996.

Remediation of uranium-contaminated soils using uranium extractants and microbial uranium reduction.
Lovley -DR, Landa -ER, Phillips -EJP, Woodward -JC, *203rd American Chemical Society (ACS) National Meeting, San Francisco, CA, 5–10 April 1992, pp. 8688-8690,* American Chemical Society, Washington, DC, 2442 p., 1992.

Table 4.2 Phase 1 Final Literature Selections (continued)

Resuspension in contaminated soils by the Chernobyl Accident [Original Title: Resuspension en suelos contaminados por el accidente de Chernobyl].
Martinez Serrano -J, Espinosa Canal -A, Aragon del Valle -A, *Radioprotection, 5: 104–115, 1997.*

Sand and organic amendment influences on soil physical properties related to turf establishment.
McCoy -EL, *Agronomy-Journal, 90(3): 411–419, 1998.*

Soil ingestion by humans: A review of history, data, and etiology with application to risk assessment of radioactively contaminated soil.
Simon -SL, *Health Physics, 74(6): 647–72,* June 1998.

Soil ingestion issues and recommendations.
Calabrese -EJ, Stanek -EJ, *Journal of Environmental Science and Health, Part A, Environmental Science and Engineering, 29(3): 517–530,* 1994.

Soil recycle and transportation model.
Hanzawa -Y, Matsuda -T, Nomura -K, *Research for Tomorrow's Transport Requirements: Proceedings of the World Conference on Transport Research, Vancouver, British Columbia, Canada, 1: 717–732,.* Vancouver Centre for Transportation Studies, Vancouver, Canada, 1986.

Soil washing physical separations test procedure - 300-FF-1 operable unit.
Westinghouse Hanford Co., Richland, WA, Funded by Department of Energy, Washington, DC, 117 p., October 8, 1993.

Statistical uncertainties in predicting plutonium dose to lung and bone from contaminated soils.
Garten -CT, Jr., *Health Physics, 39(1): 99–103,* July 1980.

Technical basis for establishing environmentally acceptable endpoints in contaminated soils.
Nakles -DV, Linz -DG, Proceedings of the SPE/EPA Exploration and Production Environmental Conference: Government and Industry Working Together to Find Cost-Effective Approaches to Protecting the Environment, Houston, TX, 27-29 March 1995, pp. 9–18, Society of Petroleum Engineers, Richardson, TX, 797 p., 1995.

Testing soil mixed with waste or recycled materials
Wasemiller -MA, Hoddinott -KB (eds.), *Proceedings of the 1997 Symposium on Testing Soil Mixed with Waste or Recycled Materials Conference, New Orleans, LA, January 16–17, 1997,* ASTM Special Technical Publication 1275, American Society of Testing and Materials, Conshohocken, PA, 327 p., September 1997.

Uncertainty and variability in human exposures to soil contaminants through home-grown food: a Monte Carlo assessment.
McKone -TE, *Risk Analysis, 14(4): 449–463,* August 1994.

Use of recycled soil for the regeneration of contaminated land.
Fleming -G, Thomson -L, *Contaminated Soil '93: Fourth International KfK/TNO Conference on Contaminated Soil, Berlin, Germany, May 3–7, 1993, Arend -F, Annokkee - GJ, Bosman -R, van den Brink - WJ (eds.), pp. 871-880,* Kluwer Academic, Boston, MA, 1993.

Utilization of fly ash for stabilization/solidification of heavy metal contaminated soils.
Dermatas - D, Meng -X, Advanced Power Assessment for Czech lignite, Task 3.6, Part 2, Sondreal -EA, Mann -MD, Weber -GW, Young -BC (eds), pp. 563–581, North Dakota University, Grand Forks, ND, 774 p., December 1995.

We're in the soils business, remember!
Toffey -WE, *BioCycle, 39(12): 57–61,* December 1998.

Whole Earth let 'em eat dirt— human and animal earth-eating behavior.
Abel -A, *Saturday Night, 113(5): 27–28,* June 1998.

4.5 Internet Searches

Acknowledging the importance of this project and the complexity of the information retrieval, NRC and NAL staffs decided to scan additional information sources. The primary purpose of the Phase 1 scanning was to give the NRC staff an overview of the availability, extent, and nature of these resources. These explorations were not exhaustive because the Internet is unstructured, diverse and ephemeral. The NRC staff agreed with this approach because of their high-priority requirement for citable sources from the literature as the primary product.

Additional information research included Internet searches using selected search engines and specific databases available via the Web, NAL networked resources, and database resources available from the library system of the University of Maryland at College Park. The NAL staff also explored and reviewed items obtained by searching the Defense Technical Information Center Web site, and statistical databases, called "Statistical MasterFile", on compact disc (CD) from Congressional Information Systems (CIS). In addition, the NAL staff reviewed titles available from the Online Computer Library Center, Inc. (OCLC) "WorldCat" international library cataloging database and conducted some searches in the Thomas Register of American Manufacturers[SM]

database on CD.

4.6 Internet Search Results

A recent report in the July 1999 issue of *Nature* noted the existence of more than 800 million indexable pages on the Web.[6] Because the volume is so great, the changes are so frequent and rapid, and the processing of complex searches for an involved topic such as this is so difficult, it was not considered reasonable to pursue the complete retrieval of all material on the Internet for this project. Additionally, the results of Internet searches do not clearly indicate the extent to which a particular item has been peer-reviewed or otherwise verified and substantiated.

Further, recent OCLC studies characterizing the Web document more than 5 million Web sites (OCLC, 2000). While the vast majority of Web sites are publicly accessible, comprehensive standards are rarely used to construct, format, or index sites, or for the search engines available to locate specific information on these sites, and the more than 800 million pages that they include.

---

[6] For more information about Internet search engines, consult <http://www.searchenginewatch.com/sereport/99/08-size.html>.

Internet coverage of specific topics is arguably as inconsistent as its rate of growth has been remarkable. Because of this lack of consistency, retrieval of subject-specific Web documents in a complex, multifaceted topic area cannot be readily planned and structured to ensure comprehensive coverage of Internet resources. Experts estimate that as many as 2,000 search engines may be available (Stanley, 1998), and each may have its own indexing systems, techniques, and methods of acquiring new sites and adding appropriate terms.

It is also noted that many search engines do not support the kind of advanced and complex search statements used in this study to retrieve items from scientific, technical, and business-related citation databases (UNN, 1999). Without truncation, proximity commands, and specific field-searching capability, the results of most Web searching for these complex concept groups, if they could be constructed and actually run, would lead to significant overflow in retrieval. If particular sets of terms and concepts can be identified and developed, some additional success might be expected for further research using general Internet search systems and techniques.

In spite of these limitations, Internet explorations using the following search engines produced several interesting and valuable documents that were submitted for NRC review:

- AllTheWeb: <http://www.alltheweb.com>

- AltaVista: <http://www.altavista.com>

- Google URL: <http://www.google.com>

- Metacrawler: <http://www.cs.washington.edu/>

Recommendations from NAL staff and external reviewers for this project also located additional resources as a result of searcher skill and experience, some good fortune, and particular experience with known sites and familiarity with particular search engines. These items were forwarded to the NRC staff in the Draft Letter

Report, Draft Letter Report Supplement, and later documents. Specific sites providing searchable database access, such as that provided by the American Society of Civil Engineers (ASCE) and searchable databases from the Defense Technical Information Center (DTIC), were located by both external reviewers and NAL staff. Although some of these may provide additional sources of published literature and other information, there is little to suggest that these items will differ from those retrieved in the exhaustive Dialog® searches. Internet Web pages and related items that were selected by the NRC staff are listed in Appendix B.

While the NRC's project needs were national in scope, pertinent local information was found on the Internet. Local and regional businesses, involved with soil as a commodity, were often listed in subject-oriented directories (i.e., for construction or landscaping) or through the local Better Business Bureau.

Although specific information on sand and gravel was not relevant to the study objectives, individuals interested in more complete material flows may explore information provided by the U.S. Geological Survey (USGS). The USGS compiles statistical reports by State for sand and gravel operations. Individuals interested in local industry should look to their own State Department of Natural Resources for more information. Ohio, for example, has provided outstanding information available on the Web at: <http://www.dnr.state.oh.us/ odnr/geo_survey/geo_fact/geo_f19/ geo_f19.htm>.

The U.S. Department of Defense has reported on its significant experience in the cleanup and remediation of former military bases. These reports may provide pertinent analogies for the NRC staff analysis of reused soil. The DTIC can provide access to much of this literature through its searchable STINET database on the Internet at: <http://www.dtic.mil>.

Similarly, the U.S. Department of Energy (DOE) has

24

pertinent experience with the cleanup and remediation of its nuclear weapons production and storage facilities. Many reports describing DOE remediation efforts were found in the National Technical Information Service (NTIS) database and reported to the NRC through this study.

Chemical contamination of soil and site cleanup are under the regulation of the U.S. Environmental Protection Agency. Several representative "Superfund" cleanup reports were provided to the NRC staff for evaluation.

4.7 Defense Technical Information Center Resources

The Internet pages maintained by the DTIC, under the Scientific and Technical Information Center (STIC) are an excellent source of defense-related and other scientific documents that have been entered into DTIC's Technical Reports Collection from late December 1974 through the present, as well as some full- text reports for those citations. This resource is searchable, using: <http://www.dtic.mil:80/stinet/>.

Searches of the STINET database, conducted by NAL staff yielded a complete listing of more than 11,000 items covering "soil(s)". Specific DTIC searches included the DOE OPENNET database which yielded a listing of about 500 titles covering soils, from which the NRC staff selected some items for further study.

One interesting item that was identified in the DTIC database search and other sources was an article published in a subject-specific issue of *Soil Science Volume 132 (1)* dated July 1981. This issue contained 18 articles reporting studies of the behavior of radionuclides in soil environments. The articles were reviewed and selected citations were added to the study results listed in Table 4.1.

4.8 Other Sources

Additional sources that were searched for this report

included the OCLC WorldCat library cataloging database, the Statistical Masterfile (SM) CD, the Thomas Register of Industrial Products, and the InfoTrac database. The first three of these sources include specific data records, and InfoTrac is a general bibliographic resource. Each of these sources is discussed below, as is NAL's information research activities.

The OCLC WorldCat database includes cataloging records for more than 40 million books and journal titles held in libraries across the United States, as well as many international libraries. NAL searched the WorldCat database to provide a sample of books that might be reviewed for possible inclusion, and to evaluate the resources for further review.

WorldCat searches for broad concept terms like "earthmoving" were tried, and several selected items were forwarded to the NRC staff for their review and selection. These items covered equipment used in this industry, in the expectation that some of this material would help to describe the physical context of exposure scenarios related to earthmoving. Other searches with terms such as "soil", "soil(s)" and "recycling", "soil(s) and material flow", "soil near5 sale(s)"[7], and "gardening and statistics" were tested as well, but these searches yielded either numerous records (over 91,000 for "soil" as a title word, and more than 120,000 as a subject term), or very few records (63 for "soil recycling", 8 for "soil(s) and material flow(s)", 9 for "soil(s) near5 sale(s)", and 54 for "gardening and statistics"). These small-yield search groups identified very few or no records with relevance for this project.

The WorldCat database may prove useful to other researchers in identifying topic-specific publications for the work of the NRC. Book titles and their subject headings are often general. However, to be effective, the complex strategies

---

[7] "near5" is a proximity operator. Terms located on either side must both be present for a record to be recalled. The terms can appear in any order and can be separated by up to five words.

that were defined for the Dialog® system database searches would need to be further refined and tailored for use in the WorldCat database system.

The Statistical Masterfile (SM) CD database includes publications with significant statistical data from U.S. Government, private, and international sources. This database was accessed at the University of Maryland's College Park library. The searches produced few relevant statistical references for this project, but some items that might be useful were noted and forwarded to the NRC. Specific search terms explored were "earthmoving" and "topsoil". Little definitive detail was found, but the search did identify items such as Pit and Quarry: State of the Industry (ISSN 0032-0293), ENR (a trade weekly for the construction industry, ISSN 0891-9526), and several documents that cover aspects of building and construction industries. Most of these related to overall industry trends and did not note the inclusion of details specific to soils. Quarries, cement, concrete, and stone data were mentioned, but not earthmoving, transport, or other uses of soil materials.

Searches in the SM database for statistics on gardening and horticulture produced listings that addressed overall production of floriculture and horticulture products. Data covering equipment such as tractors and implements was also seen, but nothing directly related to soils or soil use. Other searches using the SM CD databases focused on waste processing (including nuclear wastes). Searches were completed for topsoil, mining, minerals, and quarries, but only sand and gravel and related topics were found.

Search efforts using the Thomas Register of American Manufacturers℠ CD database provided another means to identify significant companies involved in the production of soil-related products. If additional source or producer data is required for any specific products, this would be a most convenient source of that information.

Searches using the InfoTrac system retrieved relevant citations. The NAL staff noted that these items were already included within the results of both the Dialog® comprehensive listings and the focused additional search results that were presented to the NRC in the draft letter report.

4.9 Phase 1 Observations

Extensive searching in the Dialog® system databases was the primary objective for Phase 1 of this study. The Dialog® bibliographic databases provide exceptional access to published, "citable" literature.

Unfortunately, the NAL staff determined that little systematic research or tracking has been published for parameters related to soil reuse and how humans interact with reused soil. Phase 2 of this study identified additional information resources and compiled data (see Chapter 5).

5. PHASE 2: FOCUSED SEARCH FOR PARAMETER VALUES AND RELATED INFORMATION SOURCES

5.1 Introduction

Phase 2 focused on specific parameters needed for dose analyses using codes such as RESRAD, as discussed in Section 2 and in Section 5.2, and related information sources. The specific parameters selected for study were ranked as high priority both in terms of anticipated exposure to reused soil and priority of information needed.

Phase 1 results demonstrated a lack of systematic study for parameters related to the reuse of soil and human interaction with soil. Nevertheless, a realistic understanding of these parameters was needed. As a result, the Phase 2 study extensively examined multiple sources to identify common practices, recommended guidelines, and real-life accounts. To the extent possible, inferences drawn from largely anecdotal sources were verified with secondary sources and through direct communication with experts and business representatives.

For Phase 2, a team of NAL reference specialists was convened to conduct the information research on the topics listed in Section 5.2.1. Team members initially searched online and CD-ROM bibliographic databases and the Internet. In addition, team members directly contacted experts by telephone and email when specific required information was not found in the published literature. NRC staff has significant technical expertise in the areas of civil engineering and radiological clearance. This expertise was engaged in Phase 2 to search for pertinent soil reuse information in the fields of civil engineering, construction practices, transportation, and for national and international radiological standards. The NRC staff research results are presented in Section 5.6.

The Phase 2 study was primarily directed toward identifying information sources and the presentation of typical data for in-depth examination by NRC staff and contractors conducting the dose analysis of exposure scenarios.

5.2 Data Needs for Dose Analyses

The NRC staff met with NAL researchers to explain the information needs for dose analyses and to review potential exposure pathways. These discussions emphasized the need to identify quantitative information relevant to the material flow of reused soil.

On the basis of these discussions, agricultural and landscape workers were identified as two potentially high-ranking groups for study, and further qualified specific information requirements for these groups. The added qualifications called for special attention to information concerning proximity to reused soil, working time, physical demands on the worker, coincident activities, reused soil volumes, soil depths, regional aspects and parameters, and volumes of soil additives. Table 5.1 provides an explanation of these factors.

Table 5.1 Research Factors for Dose Analyses

Factor	Consideration
Proximity	Proximity of the worker to the reused soil (e.g., hoe handle length or distance of cab seat to the reused soil) and whether the worker is in direct or indirect contact
Working Time	Hours per year that workers are in close proximity to the soil (i.e., less than 3 meters from the soil).
Intensity	The level of physical intensity (defined, where possible as moderate to strenuous) as indicated by such measures as breathing and perspiration rate
Coincident Activities	Eating and drinking on the work site, and time spent per day in contact with the reused soil
Regional aspects	Length of the growing season, hours of daylight, planting and harvesting dates
Soil Volumes	Sizes of truck beds used in transporting soil
Soil Depth	Likely depth of soil deposition or depth of tilling/cultivating and planting
Soil Additives	Volume or percentage of soil or soil mixtures in bags
Particle Sizes	Related to dust particles that may be inhaled or distributed through wind erosion processes

5.2.1 Specific NAL Research Topics

The NAL researchers initially organized the search strategy into six categories of human activities involving soil reuse. Within these categories, the NAL researchers identified the following as potentially important in the search process:

- greenhouse production and workers
  - area under greenhouse production
  - number of workers in greenhouses
  - average height of greenhouse benches
  - volume of soil used in container or bed preparation

- retail sales of bagged soil and workers
  - volume of bagged soil sold in retail outlets
  - how bagged soil is handled at the distribution points
  - number of workers employed to handle bagged soil at distribution points
  - hours per day workers are in contact with bagged soil

- farm and field workers
  - number of workers in agriculture likely to be working with hand-implements
  - days per year workers are in the field
  - ★ range of days per year
  - ★ average number of days per year
  - ★ average number of hours per day
  - depth of soil worked

- exposure of farm and field workers to dust
  - volume of dust inhaled by farm and field workers
  - particle size range of inhaled dust

29

- sources and volumes of water used for worker hydration and crop production
  - percentages of crop and fruit agriculture under irrigation
  - volume of water used in vegetable and fruit irrigation
  - sources of potable drinking water for farm and field workers
  - volume of water consumed by farm and field workers

- landscape workers
  - number of workers employed in the landscaping trades
    ★ range of hours worked per day
    ★ average hours worked per day
    ★ volume of soil purchased by the landscaping industry
  - sources of imported soil used in the landscaping industry

5.2.2 NRC Information Searches

Prior to and concurrent with the NAL research, as discussed in the previous section, the NRC staff focused on limited and relevant information sources related to reused soil. In particular, the NRC staff considered the following information categories and parameters:

- reused soil in construction practices
  - geotechnical characteristics of reused soil
  - volumes of reused soils
  - specific uses of fill materials
  - regional sources of reused soil
  - activities associated with construction use of reused soils

- transportation of reused soil
  - types of vehicles used
  - vehicle configurations (e.g., covered bed, isolated cab, and proximity of operator)
  - specific logistics of transport (e.g., duration and distances)
  - activities associated with transportation of reused soils

- landfill use and disposal of reused soil

- specific application of reused soil (e.g., cover, liner, and fill)
- geotechnical characteristics of reused soil
- volumes of reused soils
- local and State regulations for disposition
- regional sources of reused soil
- activities associated with landfill management (e.g., bulldozing)

5.2.3 International Information Searches

The NRC and NAL staffs independently considered international publications and information sources dealing with reused soil. In particular, the researchers considered the following sources:

- International Atomic Energy Agency (IAEA) Safety Standards on exemption and clearance of radiation sources and current practices

- European Commission publications on clearance and environmental restoration

- United Nations publications (e.g., UNSCEAR 2000)

- Organization for Economic and Cooperative Development/ Nuclear Energy Agency publications (e.g., recycling of materials)

- Individual foreign national efforts (e.g., United Kingdom, Department of the Environment's DETR/RAS/98.004 on Derivation of UK Unconditional Clearance Levels).

5.3 Material Flow Analysis

Since no directory of reused soil sources was identified in the study, specific information on possible origins of reused soil (Section 5.3.1) and water in contact with this reused soil (Section 5.3.2) were sought.

5.3.1 Material Flow Analysis for Soils

This Section describes possible sources, distribution, and use for reused soil. Also, commercial and professional practices regarding the composition of modern plant growth media are discussed in Sections 5.3.1.1 through 5.3.1.4.

5.3.1.1 Origin of Reused Soils

Topsoil is often sold or dispersed as a byproduct of development. The origin of the soil may be surplus material from farmland, urban, or transportation development projects. The soil may be purchased either for redistribution or in a direct transaction to the final customer. When sold for redistribution, the soil is likely to be extended and enhanced with a variety of additives (such as organic materials and fertilizers) and adjusted for pH and other physicochemical properties.

Statistics for soil transactions are largely undocumented. Presumably, this lack of documentation reflects the highly distributed nature of a low-cost resource. In addition, because a topsoil industry *per se* has not developed, few regulations and legal definitions exist with respect to topsoil (Boyles, 2000).

In order to develop an understanding about how reused soil is used a process approach was taken by NAL investigators. The process begins with the points of origin of the soil through a transaction (the point at which native soil becomes reused soil) to the final use. NAL investigators examined sources which included Federal agencies, State sponsored exchanges, and private transactions. The results of the investigation are discussed in the following Sections.

Federal

Federal mineral resources (including soil, see Table 5.2) sold in the Western United States are largely managed by the U.S. Department of the Interior, Bureau of Land Management (BLM). To sell these mineral resources, BLM uses a combination of contracts and free-use permits. A sales contract is written for a set volume of soil over a defined period of time, with upper limits ranging between 5 years for noncompetitive contracts to 10 years for competitive contracts, and multiple loads may be moved off site within the time frame of the contract.

By contrast, the vast majority of soil removed from BLM administered land is transferred via free-use permits that are issued to Federal, State, and local Governments and other non-profit organizations. (See <http://www.blm.gov/nhp/300/wo320/sndgrvl.html>.) The sale value of the BLM soil ranges from $0.20 to $0.55 per cubic yard. It should be noted BLM does not sell soil that is essential for growth of vegetation. As a followup to this study, the NAL researchers contacted several State BLM offices by telephone. Each office reported that the given State actually had very little soil to sell; they primarily sold sand, stone, and gravel to construction companies for use in road construction, parking lots, and the like.

State

Several States have established recycling services or exchanges, and topsoil is a commodity that is traded or sold via these exchanges. For example, in January 2001, the Ohio Environmental Protection Agency's Material Exchange Web site <http://www.epa.state.oh.us/opp/recyc/avail_6.html> posted two notices for raw unprocessed topsoil, indicating that volumes available for pick-up were 4,000 and 15,000 cubic yards, respectively. Both postings originated in Northern Ohio.

For example, New Jersey has established State regulations for the reuse of soils. These regulations are posted on the Internet at <http://www.state.nj.us/dep/srp/regs/soilguide/sgd53-66.pdf>. The Association of New Jersey Recyclers posts a business directory listing 10 commercial companies that sell topsoil.<http://www.anjr.com/Resources/ANJRResources/RecycledProductsGuide/Grounds&Recreation.htm>; however, as noted elsewhere in this report, what vendors describe as topsoil may in fact be recycled yard waste.

31

Private

Individuals who barter, sell, or otherwise exchange soil probably constitute the largest, most distributed, and least documented source for reused soils. The high cost of moving soil (relative to its market value) means that most reused soil is not moved far from its point of origin. State-sponsored exchanges facilitate transactions between business' and their customers. Similar exchanges have also developed in the private sector. These commercial exchanges facilitate individual-to-individual and individual-to-business transactions using the Internet as a ready, and inexpensive means to reach large audiences. One such exchange is ShopDirt.com, which posts classified ads for fill dirt, topsoil, stone, and transportation. (See Table 5.3.)

Five randomly selected business operators who had advertised on the ShopDirt.com Web site were interviewed by telephone to gather information about their soil handling operations. Table 5.4 summarizes the interview results. The business operators were also questioned about the nature of their customers and how these customers used soil-based products. (See Table 5.5.)

Table 5.2 Sale of Soil from BLM-Managed Resources in 1998 [1]

State	Contract Sales/Free-Use Permits			Production		
	Number	Quantity [2]	Value	Number	Quantity [2]	Value
Nonexclusive						
California	6	600	$ 238	6	600	$ 238
Idaho	5	271	55	5	271	55
Nevada	26	8,993	5,246	26	8,993	5,221
New Mexico	11	348	191	11	348	191
Utah	8	787	349	8	787	369
Average	11	2,200	$1,216	11	2,200	$1,215
Exclusive						
Arizona	0	0	$0	2	192	$54
California	1	300	210	0	0	0
Colorado	2	1,200	612	12	17,920	1,893
Nevada	2	1,467	750	2	1,462	748
New Mexico	11	11,288	5,248	20	17,778	4,776
Utah	1	350	88	1	350	88
Wyoming	1	57	15	0	0	0
Average	3	2,095	$989	5	5,386	$1,080
Free-Use						
Colorado	0	0	$0	1	64,900	$18,172
Idaho	2	20,500	6,150	0	0	0
Nevada	5	25,200	12,000	14	20,355	9,578
New Mexico	8	196,300	152,540	117	23,076	12,637
Utah	2	32,500	7,150	1	170	37
Wyoming	1	14,000	3,500	3	14,000	3,500
Average	3	48,083	$30,223	23	20,417	$7,321

[1] BLM, 1998 (recent data see <http://www.blm.gov/natacq/pls00/pdf/part3-20.pdf.>)

[2] Reported in cubic yards.

Table 5.3 Classified Ads for Topsoil on ShopDirt.com Web Site (11/10/2001) [1]

Location	Type	Quantity	Remarks
Glendora, CA	Fill, Topsoil	N/A	Buys and sells
London, OH	Fill, Topsoil	N/A	"A lot of nice topsoil"
Huntsville, OH	Fill, Topsoil	N/A	Farm dirt / large & small sales
Deptford, NJ	Topsoil	N/A	Buyer must transport
Jackson, NJ	Topsoil	N/A	Free, buyer must transport
Oneco, FL	Fill, Topsoil	N/A	For sale with and with out shipping
Huntington Beach, CA	Fill, Topsoil	40 yds	Free even with delivery 4-days only
DeBerry, TX	Fill, Topsoil, Clay	N/A	Buyer transports
Birmingham, AL	Fill, Topsoil	N/A	Sewer rehab 10-yr contract
Oconomowoc, WI	Topsoil	N/A	Conditioned topsoil
Waller, TX	Fill, Topsoil	N/A	Will ship outside of local area
Neptune, NJ	Fill, Topsoil	100+ Loads	$50 per load delivered within 5 miles
Hockessin, DE	Topsoil	N/A	Small and large loads, conditioned
Sarasota, FL	Fill	1 million cubic yds	Buyer has option of loading
Houston, TX	Fill, Topsoil	N/A	Company also hauls
Jackson, NJ	Topsoil	N/A	From site-work
Jackson, NJ	Topsoil	300 yds	
Jackson, NJ	Topsoil	18 yd Loads	$225 per load

[1] Listings were posted between February 26, 1999 and November 5, 2001

The interviewed companies had annual sales of soil-based products of 300 to 800,000 cubic yards. The larger operations typically sold their products to developers and the construction trade. The medium to small operations tended to have more sales to homeowners and small nurseries. For the most part, the interviewed companies sold their soil or soil-based products in bulk. None of these business operators participated in trade or professional organizations.

Unlike soils, sand, gravel, and clays are higher-value commodities with an industrial base. These commodities are tracked and annual reports are available through the U.S. Department of the Interior, U.S. Geological Survey. Current and historical summaries are available on the Internet at <http://minerals.usgs.gov/minerals/pubs/mcs/>.

Table 5.4 Telephone Interviews with Five Randomly Selected ShopDirt.com Business Owners

Category	Company				
	A	B	C	D	E
Screen soils	no	yes	no	no	no
Mixed with other components	no	yes	no	no	yes
Percentage of soil in final mixture	100%	0% 50% 100%	100%	100%	30% 50% 100%
Typical annual volume of soil product sales	800,000 cubic yds	N/A	7–800,000 cubic yds	300 cubic yds	30,000 cubic yds
Typical volume per individual sales transaction	5,000 cubic yds	5 cubic yds	10–15,000 cubic yds	5 cubic yds	6 cubic yds
Volume range for individual sales transactions	2–15,000 cubic yds	2–10 cubic yds	1–100,000 cubic yds	1–20 cubic yds	1–4,000 cubic yds
Product packaging	bulk	bulk	bulk	bulk[1]	bulk
Number of employees in contact with soil products	5	3	1	4	6
Average hours worked per day	9	9	2	8	8
Range of hours worked per day	2–12	8–10	1–8	6–10	8

[1] Some soil is bagged for sale.

Table 5.5 Customers Served and Product Uses for ShopDirt.com Soil Products

Company	Customer	Product Use
A	Grading for light commercial construction	Parking lots, fill dirt, building pads
B	Nurseries, garden centers, country clubs, homeowners, race tracks, cemeteries	Vegetable gardens, lawns, landscape, turf
C	Developers	Fill pads, building pads for drainage
D	Homeowners, contractors, landscapers	Top dressing for home lawns
E	Homeowners, contractors	Lawn maintenance

5.3.1.2 Reuse of Soil in Commercial Products

At this time, there are no Federally regulated labeling requirements for soil products other than normal requirements for weights and measures. Labeling for landscape soils and horticultural growing media is covered under the *Voluntary Uniform Product Guidelines* issued July 2001 by the Mulch and Soil Council (MSC), formerly known as the National Bark and Soil Producers Association. MSC has also implemented a product certification program, which sets standards for quality and performance. Standard labels identify product type as outlined in the Uniform Product Guideline. (See <http://www.mulchandsoil.org/industry/NB_Std_Consol_V62.pdf>.)

The relevant labeling guidelines are summarized as follows:

- Named products that refer to a single ingredient must contain 100% of the specified ingredient.

- Mixed-ingredient products with specified components must have a composition of 50% or more for the first ingredient mentioned.

- Each of the remaining individually listed ingredients in mixed-ingredient products must represent 10% or more of the product.

- Labeled mixed-ingredient products will list ingredients in descending order by percent of volume for all materials present over 2% of the final mixture.

The MSC's voluntary uniform product guidelines define the following soil mixtures:

- **Landscape Soil** is a material mixture or blend for in-ground growing of plants, which is made primarily from natural soils, bark, peat, humus, compost, and/or manure and other components.

- **Soil** is any product or material except peat or peat moss that is advertised or offered for sale or sold for primary use as a horticultural growing medium, soil amendment, or soil replacement.

- **Potting Soil** is any material for in-container growing of plants with general characteristics and productivity measures as determined with standard test protocols.

- **Premium Potting Soil** is similar to potting soil (described above) but also meets additional performance criteria.
- **Professional Potting Soil** is a formula used in commercial container production businesses with documented sales to

professional growers.

- **Topsoil** has the same composition described for "Landscape Soil," above.

Note: Other than landscape soil, there is no requirement in the guidelines for a product labeled "soil" to actually contain soil.

Examples of other international and voluntary standards for soils used in landscaping and gardening are available (Standards Australia International, Ltd., 1998; and Huinink, 1998).

5.3.1.3 Commercial and Professional Practice

Artificial planting media for the production of horticultural and ornamental crops were developed in the mid-1950s (Sheldrake, 1980). Since their introduction, the use of natural "soil" in horticulture has continued to decline (Warnicke, 1986). Components in the soil-less media include peat moss, perlite and vermiculite, various materials from agricultural or industrial wastes, and other byproducts that have been found to be helpful for horticultural production applications. In recent years, materials such as foams, gels, and polymers have been added to premium mixtures to increase water retention and enhance other performance factors.

There are many reasons to avoid the use of soils in containerized planting. Bagged materials containing soil are heavier to prepare, process, and deliver than other planting mixes. In addition, natural soils are variable, and commercial production depends upon consistent and reliable growth media. By contrast, artificial media provide growers with dependable porosity, pH, and other fertility factors. Natural soils also have biotic components and must be sterilized or treated to remove pathogens, insects, weed seeds, etc. Finally, in defined media, nutrient additions may be more readily calibrated and controlled to deliver expected fertility levels for specific applications. Lightweight non-soil materials are now the standard substrate for modern nursery and greenhouse production. (Boodley and Sheldrake, 1982).

An exception to the general practice outlined above, limited use of soils or soil materials does occur in particular situations, especially where added water-holding capacity of soil or soil components is needed, as in tropical nurseries. For example, "Soil was once the basis of most potting mixes. It is still used in nurseries in the tropics as a means of increasing water-holding capacity of media for large pots. Elsewhere, most media are soil-less" (Handreck, 1994). Growers may sometimes increase the water-supply capacity of growing media by adding clay to the potting mix (usually a particular clay mineral), rather than a composite (natural) soil material (Ehret, et al., 1998).

Table 5.6   1998 Market Share by Distributor Category for Lawn and Garden Supplies [1]

Sector	Market Share
Kmart and Wal-Mart	36%
Nurseries, garden centers, and florists	18%
National merchandisers (e.g., Sears)	14%
Home centers and discount hardware/building supply stores	12%

[1] Reported in EPM Communication, Inc., *Research Alert*, March 6, 1998. Content summarized from "The Lawn and Garden Market," from *Packaged Facts*, Scott Dempster, New York, NY.

5.3.1.4 Distribution of Soil Products

Commercially conditioned soil mixtures are sold to individuals and businesses either in packages (bags) or in bulk. Packaged (bagged) mixtures are sold to consumers at grocery and hardware stores, major retailers, and speciality nurseries or lawn and garden centers. Market leaders in lawn and garden sales are Home Depot, Wal-Mart, Kmart, and Lowe's (Howell, 2001). Table 5.6 summaries market share by distribution sector for lawn and garden suppliers. According to the Scotts Company, potting soil is the most frequently purchased item in this category. <http://www.smgnyse. com/html/growingmedia.cfm>.

The three market leaders, Scotts, Schultz, and Bayer-Pursell together account for about 85% for soil industry sales (Howell, 2001). These three companies reportedly support the MSC guidelines and standard testing for premium soils. The Scotts Company Web site reports a 49% market share for their products on the basis of a 1999 Triad Market Share Data study <http://www.smgnyse.com/html/growingme dia.cfm>. Current soil product descriptions posted for Scotts and Schultz do not identify natural soil as a component in their brand name products <http://www.schultz.com/potsoil.htm> and <http://www.scottscompany.com/ gardening/ProdGuideGarden.cfm>.

Consumer adoption of premium soil-less growth media has apparently increased dramatically in the last decade. In 1989, approximately 80% of all potting soils sold were the heavy, low-priced topsoil mixes. At the time, Hyponex (Atlanta) was the market leader with 22 warehouses (Rodgers, 1989). Presumably, these low-quality soil mixes actually included soil. The Hyponex brand is now sold by Scotts. A review of the current Hyponex product descriptions suggests that a major change in product formulation has taken place with an emphasis today on high organic content with little or no soil.

5.3.2 Material Flow Analysis for Water

Water is a carrier of environmental contaminants and another important material flow in agriculture that required analysis. Crop irrigation and livestock production are the water uses of interest for this study.

5.3.2.1 Water Use in Crop Irrigation

Crop irrigation returns significant benefits to farmers through increased yields. In 1998, the USDA reported average yields of 163 bushels per acre for all irrigated crops across the country, compared to an average yield of 114 bushels per acre for non-irrigated crops (USDA, National Agricultural Statistics Service, 1998, Table 22).

Yield alone is not the only factor used in deciding when and how to irrigate. Environmental factors like temperature, humidity, and wind speed are just as important as economic and property factors such as, price and availability of water. Given the complexity of this decision-making process, producers relied most on the condition of their crops or the feel of the soil. External factors (such as scheduled water delivery) was the third most commonly cited decision-making method. In a small number of cases, producers used computer simulation models (USDA, National Agricultural Statistics Service, 1998, Table 29).

Irrigation is the predominate use of water in the Western United States (Solley, et al., 1998). In 1995, for example, water withdrawn for irrigation was estimated at 134,000 million gallons per day, representing 39% of freshwater use for all offstream categories. Surface water was the source for 63% of the water used in irrigation.

Ground water pumped from wells is another significant source of water for irrigation. In 1998, nearly all of the farms with functional wells used them for irrigation that year. (See Table 5.7.)

California alone accounts for 17.4% of the irrigated land in the contiguous United States. Texas accounts for an additional 11%, Nebraska 13%, and Florida 3.2% of the irrigated land. Nearly 100% of the harvested acres used to grow rice were irrigated. Similarly, 35 to 70% of the harvested acres used for orchards, Irish potatoes, vegetables, berries, sugar beets, dry edible beans, sugarcane, and cotton were irrigated (Moore, et al., 1997).

Table 5.7 Irrigation Wells on Farms [1]

Category	Number
All farms with capable wells	91,500
Number of capable wells	374,072
Number of farms using wells	85,014
Number of wells used	336,040

[1] Extracted from Table 12, Irrigation Wells on Farms: 1998 & 1994; excludes abnormal and horticultural speciality farms (1998 Farm & Ranch Irrigation Survey, Census of Agriculture).

Available soil moisture (ASM), an important determinant for irrigation, varies by soil type and the amount of water in the soil. Many crops (including vegetables) are sensitive to drought damage. A North Carolina Extension publication recommends applying up to 1.5 inches of water each week during hot periods for plants with a surface spread of more than 12 inches. This irrigation level can be decreased to 0.75 inches per week in cooler seasons. In general, application rates should not exceed 0.4 inch per hour for sandy soils, 0.3 inch per hour for loamy soils, and 0.2 inch per hour for clay soils (Sanders, 1997).

Each crop has critical developmental stages and preferred minimum soil moisture levels dictating irrigation needs for crop production. See Table 5.8 for the recommended irrigation rates for selected vegetable crops (Sanders, 1997).

5.3.2.2 Water Use for Livestock Production

In 1995, water withdrawn for livestock production was estimated at 5,490 million gallons per day, representing nearly 2% of freshwater use for all offstream categories. Surface water was the source for 59% of the water used for livestock production. This category of water use increased by 22% between 1990 and 1995, largely because of expanded production of fish raised in captivity. By State, Idaho consumes the largest volume for total livestock production (Solley, et al., 1998).

Table 5.8 Vegetable Irrigation Needs [1]

Crop	ASM [2]	Inches/Days
Irish potato	70%	1/7
Tomato	50%	1/5-7
Beets	20%	1/14
Carrot	50%	1/21
Edible Soy	40%	1/14
Cantaloupe	60%	1/10
Onion	70%	1/7
Lettuce	60%	1/7

[1] Extracted from Sanders, 1997
<http://www.ces.ncsu.edu/depts/hort/hil/hil-33-e.html>
[2] Available soil moisture is the percent of soil water between field capacity and permanent wilting point.

5.4 Characterization of the Green Industry

Texas, California, and Florida were selected for more thorough analysis in this study. These States were selected for their importance in agricultural production, the diversity and length of their growing seasons, and their high population densities.

The Texas Nursery & Landscape Association (TNLA) published a major study of the economic impact of the green industry in Texas for the year 2000 (Hall, 2001). The green industry includes allied input suppliers, wholesale growers, retail garden centers, and landscape firms. Home centers and mass merchants also represent a significant portion of green industry retail sales. Table 5.9 summarizes relevant labor statistics from the TNLA report.

Independent validation of the TNLA study and further characterization of soil transactions by mass merchants was sought through contact with local businesses. Three Maryland-based mass merchants were contacted by telephone to review their soils-based business activities. Table 5.10 presents the results of these interviews. Store brand topsoil and landscape soils sold by mass merchants may include reused soils.

Table 5.9 Texas Green Industry Labor Force Statistics [1]

Sector	Number of Firms	Number of Employees	Average # Employees	Retail Market Share
Home Centers and Mass Merchants	715	83,292	†	38% (Hm Cntr) 30% (Mass Mer)
Retail Garden Centers	3,464	39,196	23	32%
Nursery Growers	997	19,325	19	
Landscape Firms	11,951	80,820	42	

[1] Extracted from Hall, 2001 (with the Texas Nursery & Landscape Association).
† Lawn and garden departments of home centers (such as Home Depot) and mass merchants (such as Wal-Mart) employ about 33% of the green industry labor force. Firms in all sectors expect a 30% increase in green industry jobs over the next 5 years.

Table 5.10 Maryland-Based Mass Merchant Soil-Based Business Activities

Business Activity	Company		
	A	B	C
Source of local or generic soils	N/A	Pennsylvania	Delaware
Reported composition of soil products	80% Topsoil	Ultra Brand a blended product & plain topsoil	Topsoil
Typical annual sales of soil products	12,000 40-lb bags	80,000 40-lb bags	11,000 40-lb bags
Typical volume per individual sale	6–10 bags	10 bags	8 bags
Range of volumes sold per individual sale	1–60 bags	1–60 bags	1–60 bags
Type of customer	Individuals	Homeowners	Homeowners
Number of employees in contact with soil	5	20	5
Average number of hours worked per day [1]	7–8	8	8
Range of hours worked per day [2]	4–10	8–14	6–8

[1] Across all employees
[2] Work shift range for individual employees

These merchants distribute their products primarily to individuals. Individuals purchasing these soil products appear to use the products for indoor plantings and, outdoor flowerbeds and gardens. One business representative felt that a small number (1 out of 60) were using the soil products for lawn improvements. Another business representative also recognized that individuals were using their soil products for planting, but noted that the store did not recommend the product for that application. This same representative mentioned other more appropriate uses such as filling holes, fixing erosion damage, and leveling ground before planting lawn seed. A representative from another company mentioned that their soil was also used in planting trees, shrubs, and grass.

5.4.1 Greenhouse Practices and Parameters

Containerized plant culture is a significant production practice in greenhouses and could be considered a significant end-use for reused soils. Section 5.4.1 discusses what is known about the composition of plant growth media used, food crops grown, and general labor statistics in greenhouse production.

5.4.1.1 National Greenhouse Production

The U.S. vegetable greenhouse industry is a mixture of small family-run operations (2,500 to 10,000 square feet) and a small number of larger multi-acre operations (10 or more acres) (Greer and Diver, 2000). Tomatoes are the leading greenhouse commodity, followed by European cucumbers, lettuce, peppers, and culinary herbs. California is the leading State in greenhouse production, followed by Florida, Colorado, Arizona, Ohio, Texas, and Pennsylvania, each with over a million square feet.

Soil-based greenhouse production is mentioned in the Appropriate Technology Transfer for Rural Areas (ATTRA) publication, entitled "Organic Greenhouse Vegetable Production." The publication does not offer detailed data for the extent of this production, but it does note that 40% of greenhouse acreage was soil-based in 1988, dropping from 70% in 1974. A literature review suggests that soil-based greenhouse operations will use a substrate that is not likely to contain more than one-third actual "soil," leaving a very small percentage of greenhouse production that is directly involved with soil.

This observation is generally confirmed in a communication with a California-based county extension specialist, who made the following statement:

> "The greenhouse veg statistics are actually not easy to come by. There is no national association that compiles this, other than the every 10 years or so Census of Agriculture. And even then, there is usually no distinction between soil and soil-less culture. I can tell you that almost all the major U.S. producers grow in soil-less culture. Many of the small, 4000 sq. ft or less, operations do grow in the soil."

The USDA's National Agricultural Statistics Service (NASS) gathers an extensive range of statistical information about many aspects of the agricultural sector. Regular surveys are conducted every 5 years for the Census of Agriculture. (The Census is a complete accounting of agricultural production for all operations that would normally expect annual sales of $1,000 or more.) NASS also works in collaboration with the Bureau of Labor Statistics (BLS). Table 5.11 summarizes NASS 1998 data for greenhouse food crop production in selected States.

Table 5.11   1998 Greenhouse Food Crop Production in Selected States [1]

State	Number of Employees [2]	Number of Operations	Area (x 1,000 sq. ft.)	Total Sales (x $1,000)
Total U.S. Production	215,080	1,015	31,644	222,624
California	137,980	97	9,789	69,027
Florida	16,680	35	3,154	17,169
Colorado	1,370	22	4,113	35,257
Arizona	8,670	11	D [3]	D [3]
Ohio	1,410	42	829	3,356
Texas	7,340	32	1,475	5,886
Pennsylvania	1,050	68	727	10,009

[1] Extracted from USDA, 1998 Census of Horticultural Specialties
<http://www.nass.usda.gov/census/census97/horticulture/table37.pdf>
[2] Extracted by State from the 1999 BLS Web site <www.bls.gov> in the labor category, farm workers, and laborers, crop, nursery, and greenhouse
[3] Data withheld to avoid disclosing information about individual operations

5.4.1.2 Distance Factors

Greenhouse agricultural production commonly refers to a broad spectrum of covered structures, ranging from a total glass enclosure (building) to a pole-supported plastic film covering field plants, or a shade or temporary cover. Plants grown under various cover will either be planted directly in the ground, in raised beds, or in container pots on greenhouse benches. Organic farmers in Canada are raising tomatoes in 1-foot-high raised beds (Greer and Diver, 2000). Most greenhouse benches are generally 30 to 36 inches tall. Plants grown using container culture, hydroponic, or other soil-less production methods are usually placed on benches.

5.4.2 Landscape Trade Practices and Parameters

Landscapers are professionals who purchase, transport, and deposit soil and soil-like materials in their work. Their work brings them into close contact with reused soil. Section 5.4.2 discusses professional practices, labor statistics, commonly used equipment, sources of reused soil, and sod production.

5.4.2.1 Landscape Practices

Nearly 40% of all agricultural service workers (including landscapers) are employed in California, Florida, and Texas. Other States employing significant numbers of landscape workers are Arizona, Illinois, Pennsylvania, Ohio, and New York. According to the BLS, landscape workers typically are involved in mowing grass; planting, watering, and pruning trees and bushes; mulching; preventative pest spraying; etc. Landscape work is often dependant upon regional season, weather factors, and amount of daylight available during an average work day.

For most of the contiguous United States, landscape employment is seasonal and part-time. Landscape workers often work from dawn to dusk, 6 to 7 days during any given week, from Spring through Fall (BLS, 1997). Nationally, landscaping, grounds-keeping, nursery, greenhouse, and lawn service workers held about

1,285,000 jobs in 1998. Employment was distributed as follows:

Landscaping and groundskeeping
laborers...130,000
Lawn service managers...................86,000
Pruners...45,000
Sprayers and applicators................19,000
Nursery and greenhouse
managers..5,000

About one-third of wage and salaried workers were employed in companies providing landscape and horticultural services. Others worked for firms operating and constructing real estate, amusement, and recreational facilities (such as golf courses and race tracks), and retail nurseries and garden stores. Some were employed by local governments, installing and maintaining landscaping for parks, schools, hospitals, and other public facilities (BLS, 1997).

It should be noted that undocumented immigrant workers may confound the accuracy of conventionally compiled labor data for landscape workers. Up to 52% of farm and field workers in California are undocumented (Furillo, 2001). This problem is of special concern in California, Texas, and Florida, which have high populations of undocumented immigrants and which employ a large number of landscape workers.

Handbooks for time and job estimates provide practical information by task. These handbooks are potentially useful in estimating realistic soil contact time in landscaping and groundskeeping activities (Nilsson, 1996).

5.4.2.2 Commonly Used Equipment in Landscaping

The tools and machinery used in soil-related tasks vary widely, depending on the task being performed, the scale and nature of the landscape site, and the size and skills of the work crew. Activities involving imported soil include loading, delivering, and unloading bulk and bagged soil and soil-like materials; spreading, grading, and leveling delivered materials; incorporating imported materials with onsite soil; handling materials used in installation of hardscape items (pathways, pools, etc.) and plant material; and topdressing and aerating established turf and garden areas. Specialized applications include golf course and tennis court construction and maintenance and leveling of athletic fields.

Large-scale projects may involve front-end loaders, dump trucks, skid loaders, bulldozers, backhoes, graders, large tractors, and tractor-pulled attachments (such as tillers and spreaders, trenchers, and tree spades). Smaller jobs and job sites typically rely on pickup trucks, ride-on type tractors with appropriate attachments, rototillers, drop and rotary spreaders, sod rollers, long-handled hand tools, and, in some cases, trowels and short-handled weeders.

Selected Large-Scale Equipment Used in Landscaping Tasks

- **Large-scale loading and transporting** equipment include a dump truck, articulated dump truck, pickup truck, front-end loader, loader, four-wheel loader, backhoe loader, bucket, skid loader, skid steer loader, skid loader attachments, or related loader and tractor attachments (such as a blade, breaker, forks, pallet forks, tree spade).

- **Grading and leveling** equipment may include a earth-moving or excavation equipment, as well as an excavator, mini-excavator, earth compacter, grader, motor grader, bulldozer, dozer scraper, tamper, trencher, tractor, compact tractor, backhoe, or related tractor attachments (such as a blade, scraper or box scraper).

- **Spreading and incorporation** equipment may include a spreader, drop spreader, rotary spreader, wheelbarrow, rake, york rake, harley rake, broom, scarifier, cultipack, tiller, rototiller, rotary hoe, rake, plow; or related tractor attachments (such as a blade, breaker, cultivator, disk, harrow, scarifier, or tooth bar).

- **Plant installation** equipment may include a backhoe, shovel, spade, trowel, or tree spade.

- **Hardscape installation** equipment may include a backhoe, post hole auger/digger, shovel, spade or other equipment needed to install irrigation systems.

- **Turf maintenance** equipment may include an aerator, corer, vertical mower, thatcher, dethatcher, edger, or specialized equipment used at athletic facilities.

Landscape contractors typically use light- or medium weight trucks from 10,000 to 19,000 pound weight with a hauling capacity of 3-4 tons (2.7 – 3.6 metric tons) (Wessling, accessed January 30, 2002). One South Carolina firm hauls truck loads of 15 to 16 tons (13.5–14.4 metric tons) (Deese, Landscaping, Hauling, and Grading, 2001). Additional information and equipment recommendations for landscaping can be found at <http://www. igin.com/Landscaping/index.html>.

5.4.2.3 Soil Material Sources in

Landscaping

Today, many factors impact selection decisions for soil and soil amendments in the landscaping industry. Among these factors are State and local laws that reduce the amount and nature of materials that can be deposited into landfills. Environmental regulations mandate nutrient management plans with important restrictions on phosphorus loading for soils. Fertility and soil characteristics are often not optimal for plant growth, and apparently soil is not always readily or reliably available. The combination of these factors is encouraging a trend away from the use of soil in the landscaping industry.

Despite the observed trend, some soils continue to be used in the landscape industry. The sources of these soils appear to be largely local and highly distributed. As a result, the source, volume, blended mixture composition, and handling of these soils are not tracked by professional or Governmental bodies. Anecdotal information provides insight for consideration. For example, one Ohio-based landscape supply firm provides component information for its soil-based products, and is summarized in Table 5.12.

Table 5.12 Landscape Soil Blends from an Ohio Supplier [1]

Product Name	Composition
Topsoil	60% topsoil, 20% organic compost or humus, and 20% coarse sand, shredded, pulverized, and screened through a ¾-inch screen
Econ-Blend Soil	Approximately 70% soil, 15% compost, and 15% sand, screened through a ¾-inch screen
Mix Soil	40% topsoil, 30% organic compost or humus, and 30% coarse sand, shredded, pulverized, and screened through ¾-inch screen
Planting Bed Mix	30% topsoil, 30% organic compost or humus, 20% coarse sand, and 20% pea sized silica gravel, shredded, pulverized, and screened through a ¾-inch screen
Topdressing Mix	50% course sand and, 50% organic compost or humus, screened through a ½-inch screen.
Commercial Soil	100% upland topsoil, screened through a 1-inch screen.
Virgin Soil	75% natural topsoil, 25% organic compost or humus, not shredded or pulverized, but screened through a 2 inch screen.
Fill Dirt	Primarily clay

[1] Recommended uses for these formulations are available on the company's Web site at <http://www.three-z.com>.

5.4.2.4 Commercial Sod Production

Current sod production and harvesting practices cause little soil depletion. The small amount of soil that is lost in each sod harvest is somewhat variable, and differences are mostly attributable to the variety of grass species harvested. Sod production is generally seen as an activity that builds soil rather than depleting soil. Wind and water erosion of farmland soil is significantly greater in other agricultural commodities such as wheat, cotton, and corn (McCarty, 1994). Even the Internal Revenue Service (Ruling 79-267, issued September 1979) disallows land depreciation allowances for soil depletion in sod operations, (Turfgrass Producers, 1995).

In producing sod, activities centered around leveling and preparing the seedbed for planting are the processes most likely to involve substantial direct contact with soil. Once the fields have been leveled, the soil is usually compacted to firm the seedbed.

Sod is harvested by cutting through the grass root zone. This root zone is composed of organic material from the plant roots and a thin layer of mostly organic material. Usually, soil is not added back to the harvested fields. This observation was confirmed through telephone contacts made by NAL investigators with three large commercial producers.

Turfgrass is harvested in several ways. Ideally only ¼ to ½ inch (0.6 to 1.3 cm) of root zone should be removed when sod is cut (McCarty, et al., 1999). Large operators use mechanical cutters and harvest strips that are 12 to 16 inches (30.5 to 40.6 cm) wide by 24 to 36 inches (61 to 91 cm) long; the tractor-mounted or self-propelled harvesters reap 600 to 800 square yards per hour. Small operators typically use a small, hand-operated, walk-behind unit and harvest 150 to 200 square yards per hour (125.4 to 167.2 square meters per hour).

Sod is either rolled or stacked in flat sheets on pallets. Approximately 400 to 500 square feet of sod is stacked per pallet, and a forklift is then used to load the pallets onto trucks for shipping. The usual tractor-trailer load consists of 10,000 square feet of sod. Newer "big roll" harvesting methods cut continuous strips that are 42 inches wide and up to 100 feet long. With this method, each tractor-trailer carries twenty-four 100-foot rolls.

Sod growers can harvest up to 40,000 square feet per acre per cutting. This represents the upper end; more typically, growers will cut 28,000 to 38,000 square feet per acre (McCarty, 1994).

5.5 Agricultural Practices and Parameters

Section 5.5 provides information about the farm and field work force and the various environmental pathways (such as dust exposures, tillage depths, and water consumption) by which these workers might be exposed to the effects of reused soils.

5.5.1 Farm and Field Workers

The National Agricultural Statistics Service (NASS), the Economic Research Service (ERS), and the U.S. Department of Labor collect statistics on farm labor. On a quarterly basis, NASS publishes the Farm Labor Survey, which reports data on the average number of hours worked per week by region for the United States as a whole. The 18 regions average 1 to 3 States per region (see Table 5.13 for selected data). The reports are available online at <http://usda.mannlib.cornell.edu/reports/nassr/other/pfl-bb/>. The ERS Web site also has a "farm labor briefing room" that provides links to a number of important statistical resources at <http://www.ers.usda.gov/briefing/FarmLabor/farmlabor/>. Finally, the Department of Labor compiles the National Agricultural Workers Survey (NAWS), which provides more detailed information regarding the working and living conditions of farm workers. The survey is available at <http://www.dol.gov/asp/programs/agworker/naws.htm>.

The *Sacramento Bee* also publishes information about California farm workers, and its Web site documents labor and pay abuses and other farm related stories. On May 21, 2001, the *Bee* published a special report on farm labor statistics. It should be noted that this survey reported 52% of farm and field workers were undocumented in California.

The usual planting and harvesting dates for agricultural commodities are another data source to identify seasonal work periods, which vary by commodity and geographical location. NASS publishes Agriculture Handbook Number 628, "Usual Planting and Harvesting Dates for U.S. Field Crops," which reports dates by State and field crop. The publication is available online at <http://usda.mannlib.cornell.edu/reports/nassr/field/planting/>.

Table 5.13. Number of Farm Workers Employed and Hours Worked Per Week [1]

Item	Florida	California	Texas & Oklahoma	Total United States
All hired workers [2]				
Jan 7–13, 2001	55	190	65	678
Oct 8–14, 2000	50	242	61	952
Jan 9–15, 2000	60	204	44	685
Average hours				
Jan 7–13, 2001	37.6	35.7	38.1	36.9
Oct 8–14, 2000	38.9	43.2	37.3	41.2
Jan 9–15, 2000	41.9	42.7	37.6	38.4

[1] Extracted from the Florida Agriculture *Farm Labor* Report, a joint publication of the USDA, NASS, the Florida Department of Agriculture and Consumer Services, and the University of Florida, Institute of Food and Agricultural Sciences, issued February 21, 2001.
[2] Reported in thousands

5.5.2 Tillage Depth

The depth to which soil is prepared for planting varies by plant variety, soil condition, and production practices. Tree root balls, for example, are typically placed into holes that are at least 6 inches deeper than the expanse of the root ball (which can reach several feet in diameter), while lettuce seeds are spread on the soil surface with a very light covering of soil or compost. Typical tillage depths recommended to enhance soil conservation are provided by the Natural Resources Conservation Service in its "CORE4 Conservation Practices Training Guide: The Common Sense Approach to Natural Resource Conservation." Some general guidelines from the training guide include:

Chisel plows	4 to 8 inches
Disk harrow primary cutting	4 to 8 inches
Disk harrow finishing	2 to 6 inches
Moldboard plow	4 to 8 inches
Row planters	1 to 2 inches
Row cultivators	1 to 3 inches

General information about the CORE4 conservation program is available on the Internet at <http://www.ctic.purdue.edu/Core4/Core4Main.html.>

Recommended planting depths for corn varies from 1 to 3 inches, depending upon the planting date, soil moisture, and temperature, among other factors. A good average planting depth is 1.5 inches, as recommended by the Purdue University Extension Service at <http://www.agry.purdue.edu/ext/corn/rln-bio.htm>.

5.5.3 Farm and Field Worker Exposure to Dust

Any activity that rips the land generates a range of particles that can become airborne. Some of these particles are of sizes and shapes that pose inhalation health risks to humans. Particles in the size range (i.e., smaller than 4 µm in diameter) that can deeply penetrate the lung cavity are of particular concern. These particles are commonly referred to as respirable dust (RD). Clausnitzer and Singer (1997) studied the concentration and mineral content of RD generated by agricultural practices in California. They concluded that corn and tomato production were the most intensively cultivated crops in their study and the crops that generated the greatest concentrations of RD (see Table 5.14).

The source materials for RD generated through agricultural production practices were further studied by Clausnitzer and Singer (1999), who determined that the mineral composition of the RD paralleled that found in the source soil. In this study, the authors also analyzed particle sizes and shapes and found that approximately 60% of the sampled dust particles generated in soil ripping operations were under 4 µm. (Clausnitzer and Singer, 1999).

Table 5.14 Average Respirable Dust Concentrations from Common Farming Practices[1]

Farming Operation/Source	Dust Concentration (mg/m³ air)
First finish disking	3.788
Second & third finish disking	4.936
Land planing	13.604
Disking wheat stubble into soil	7.158
Tomato harvest 75 cm above surface	3.681
Ripping first and second	9.885
Plowing	6.463
Corn harvest	6.688
[1] Excerpted from Clausnitzer and Singer, 1997.	

5.5.4 Water Consumption by Farm and Field Workers

Water use involving surface water and/or ground water that have contacted the reused soil is an important environmental pathway. Consequently, the researchers in this study sought information on water ingestion and other contacts with water. Agricultural work is primarily conducted outdoors and frequently when ambient temperatures are high. Subsequently, agricultural workers are at high-risk for heat-related illnesses. This condition is usually prevented by increased water consumption. Data on water use by agricultural workers at high risk heat-related illnesses were identified. (See <http://www.cdc.gov/niosh/ nasd /docs2/nj00800.html>.)

The U.S. Military and the National Institute for Occupational Health and Safety (NIOSH) have issued guidelines for water replacement, recommending that workers drink 24 oz/hr during moderate work when the temperature is 82–90° F. Water replacement rates should increase to 33 oz/hr for more strenuous activity or when ambient temperatures are 90° F. The guidelines further recommend replacing water frequently in small amounts and resting every hour. (See

<http://www.capnhq.gov/nhq/cp/ encampments/AETC.htm> for further detail.)

National and State regulations have been established to protect the health and safety of farm and field workers. The Occupational Safety and Health Administration (OSHA) directs employers with 11 or more employees to provide toilets, potable water, and hand-washing facilities for workers engaged in "hand labor operations in the field." (See <http://www.osha-slc.gov/OshDoc/ Fact_data/FSNO92-25.html>.)

It is difficult to track compliance with the regulations. For the period from 1996 through1998, 39 violations were recorded for the entire State of California. The Director of the Virginia Commonwealth's Occupational Health Compliance Division reported that violations were generally recorded only after a complaint was lodged or an illness occurred. On the other hand, the National Agricultural Workers Survey (NAWS) for 1997–1998 reported that 98% of the surveyed workers had access to drinking water, but 16% reported not having water to wash their hands. (See <http://www.dol.gov/asp/programs/agworker / report_8.pdf>.)

Agricultural worker hydration is under study at the University of California at Berkeley <http://are.berkeley.edu/APMP/>. To date, the university conducted two small field trials <http://are.berkeley.edu/heat/heat2/trial1.ht ml#Findings>, which indicated that the actual consumption of water by workers is influenced by several factors. Basis of pay was one factor; workers who were paid on a "piece work" basis were less likely to stop for water breaks than those who were paid on an hourly basis. Convenience of access to the coolers was also important in how frequently the workers stopped for water.

Climatic conditions affecting the harvest dramatically affected worker activities, including the number of hours worked and the amount of water consumed. Finally, the field trial generally observed that workers stopped for water fewer times and drank larger volumes than recommended in the guidelines for optimal hydration. (See <http://are.berkeley.edu/ heat/> and <http://are.berkeley.edu/heat/ heat2/>.)

Even less information is available regarding the sources of drinking water provided to farm and field workers. According to the U.S. Environmental Protection Agency (EPA), groundwater is the source of drinking water for 53% of all Americans. (See <http://www.epa.gov/safewater/ wot/wheredoes.html>.) Groundwater is the more usual source of drinking water in rural areas. (See <http://www.epa.gov/ safewater/dwh/where.html>.)

Workers providing their own water are likely to obtain that water from their homes. Therefore, it is important to know the proximity of the worker's home to the fields. Beginning in December 1997, the Housing Assistance Council (HAC) conducted a 10–month survey of farm worker housing for the Eastern migrant worker stream. The survey collected information on 1,566 units housing 8,965 people. Commonly, 39% of these units were directly adjacent to farm fields (although it varied from 11–93%). The usual length of stay in these units was 5.3 months (although it varied from 2.5 to 6.7 months with the longer stays predominate in southern States such as Florida) (see <http://www.ruralhome.org/ pubs/farm worker/meager/toc.htm#table>).

The NAWS survey found that up to 28% of all farm and field workers use housing provided either free of charge or rented through their employers. The smaller HAC survey found that 55% of farm worker housing units were owned by the employers. A large percentage of these housing units are adjacent to farm fields (63–88%).

5.5.5 Urban Food Production Practices

People are moving out of rural regions and into cities. This year more than 50% of the world's population lives in metropolitan areas. As cities grow, they tend to sprawl outward from the city center to overtake productive farmland. These trends together increase food costs to consumers.

Responding to the need for an affordable food supply, many community action groups are working to develop "sustainable cities." Sustainable cities supply their citizens with locally produced food. This food is largely grown using organic production principles, which focus on waste reduction and non-chemical solutions. Organic production emphasizes soil building through recycling yard and food residue, and other municipal wastes.

Urban agriculture is not limited to the inner city. Gardening for food production is an activity enjoyed by many. Nugent, 1997, noted that in 1991, urban areas accounted for an estimated 33% of all crop and livestock sales and involved approximately 25% of all households. New York City alone has more than 1,000 community gardens

(Canada, Office of Urban Agriculture, 1997).

Urban agricultural practices are more diverse than rural operations. Metropolitan food production takes place in container pots on rooftops or window sills, abandoned city lots, backyards, community gardens, recreational farms, and adaptive farms, as well as traditional farming operations.

Land is often a limiting resource in urban settings and, as a result, the land is more intensively cultivated. U.S. urban farms sell 13 times more per acre than non-urban farms. Local restaurants purchase a high percentage of their produce from local sources (Canada, International Development Research Centre, 1994, and Heimlich and Barnard, 1992).

Beginning in May 2000, Canada's Office of Urban Agriculture posted a 1-month survey on the *City Farmer* Web site. The survey asked questions about the nature of and interests in urban food production practices. The 100 responders from 16 countries provide information on the realistic practices of food production in metropolitan settings. Table 5.15 summarizes the results of the survey. The Canadian survey is supported by similar data from the 1996–1997 National Gardening Survey, conducted by the Gallup Organization, Inc. on behalf of the National Gardening Association (Butterfield 1997).

Table 5.15 Urban Agricultural Practices [1]

Category	Response	% of Respondents
Urban food production		81%
Size of production area	0–100 sq ft 100–1,000 sq ft 1–10,000 sq ft > 10,000 sq ft	35% 37% 10% 3%
Location of production area	Home gardens Community gardens Window Sills Balconies Other [2]	35% 21% 14% 10% —
Product grown	Vegetables Herbs Fruit Legumes	77% 72% 45% 36%
Country of Origin	United States Canada 14 other countries (combined)	56% 26% —

[1] Results of Canada's Office of Urban Agriculture Survey, conducted in May 2000, excerpted from *City Farmer*.
[2] Including rooftops, school gardens, parking lots, vacant lots, etc.

In 1998, the American Community Gardening Association conducted a study on the status of community gardens. The respondents included more than 6,000 representatives from 38 cities. Only 1.5% of the gardens were "owned" or in a permanent land trust. Neighborhood gardens predominated (67.4%), followed by public housing gardens (16.3%, approximately 978, of which 834 were in New York), and school gardens (8.2%). Gardens administered by senior, mental health, rehabilitation, and economic job center gardens together made up the balance (3.4%). Community gardens had grown in number by approximately 30% in the 5 years preceding the survey (Monroe-Santos, 1998).

A survey conducted in 11 Latin American countries found that urban agriculture could not fully replace other work to support family needs. Working 1 to 1.5 days per week was required to maintain the urban garden for the average Latin American family (Nugent, 1997).

Raised beds are often used for intensive, small-scale gardening. The "Journey to Forever" Web site recommends digging trenches to prepare a "deep growth zone" for the roots of plants growing in raised beds. The bed is constructed by first digging a trench to a depth of 16 inches and mixing 4 inches of compost with the soil taken from the trench. Walls are built to hold the beds to a height of 12–15 inches. Beds are recommended to be no more than 30 inches wide, but preferably 24 inches in width and 8 feet long with a 15-inch path around the beds. Local municipalities and agricultural extension offices were listed as sources for compost to be used in place of added soil. <http://journeytoforever.org/garden_sqft.html>

5.5.6 Additional Information Sources from Technical Peer Reviewers

NRC staff reviewed the Phase 2 technical peer reviewers' comments for identifying additional information sources. One technical reviewer discussed the U.S. Environmental Protection Agency's (EPA)
technical support documents for biosolids, (known as the "503 Biosolids Rule," EPA, 1994 available at <http://www.epa.gov/owm/ bio/503pe/>). Technical assessments for the "503 Biosolids Rule," are provided in EPA, 1995 and is available at <http://www.epa.gov/owm/bio/503rule/503gtoc.pdf>. The NRC staff has reviewed this work as part of its cooperative efforts in ISCORS project to develop technical guidance on sewage sludge. This ISCORS project is also assessing land application scenarios, and the NRC staff has reflected this coordination with the NRC staff soil reuse study.

Additional comments focused on information related to soil ingestion and garden foods. These references are: Chaney and Ryan, 1993; Ryan and Chaney, 1993; Chaney and Ryan, 1994; and Chaney et al., 2001. In the context of soil removal and reuse, the NRC staff views soil ingestion as a secondary human pathway and will consider this information in the scenario characterization and modeling analysis.

Further information sources on this topic focused on the measurement of soil ingestion by children and sources of error in such estimations. These references are Stanek et al., 2001 and Cohen et al., 1998.

5.6 NRC Staff Research and Results

The NRC staff utilized the Phase 1 literature findings to search for additional information sources in the areas of construction activities, transportation, international studies, and national radioactivity standards. The search process involved reviewing professional journals and documents promulgated by the International Atomic Energy Agency (IAEA), Nuclear Energy Agency (NEA), and various Federal agencies including documents prepared by national laboratories operated by the U.S. Department of Energy for information sources related to soil reuse activities and dose assessment studies. Table 5.16 lists specific literature citations and Web sites identified in the NRC staff Phase 2 search.

5.6.1 Construction Activities Involving Reused Soil

The engineering properties of soil materials used in the construction industry may vary, but are often improved by selection, compaction, moisture control and mixing. In particular, the types of soil materials that may be specified for construction include dumped fill, selected fill, blended fill, modified fill, hydraulic fill, and backfill. Each application requires specific materials and conditions, as described in the following paragraphs.

One of the oldest building materials still in use today as the primary building material in many parts of the world and to some extent in arid regions of the United states is ordinary soil. There are several types of building practices using earthen materials, the most common being adobe and rammed earth construction.

5.6.1.1 Fill Materials

Dumped fill involves moving soil materials from an excavation site, and depositing those materials to fill an area to surveyed lines and grades, as in the construction of roadways, canals, land reclamation projects, covers for landfills, and dams. Dumped fills generally must be free of tree stumps, organic matter, trash, sod, peat and other such materials. However, in the reclamation of wet lands, swamps, or water fronts, dumped fill may include construction waste (e.g., broken up concrete, bricks, metal scrap, and similar incompressible materials).

Select fill includes selected sands, gravels, rockfill or rip-rap, and impervious fills (clays/silts) for various specific engineering purposes. Blended fills contain two or more materials that by themselves do not have adequate engineering properties but when combined together produce a satisfactory material. Such materials may be used as sub-bases or bases for roadways or the construction of dams, and so forth.

Modified fills are fabricated by adding minute quantities of selected admixture (such as lime, cement, or asphalt). Hydraulic fills involve the placement of fill materials under conditions of excess water content. These situations may involve excavating and transporting fill using flowing water or placing fill in still water.

Backfills are used to refill around structures in confined spaces. Backfills may be compacted clayey or silty soils or compacted cohesionless free-draining soils (U.S. Department of the Interior, Bureau of Reclamation, 1990).

5.6.1.2 Soil as Building Material

Soil can be used to fabricate building materials such as adobe, or may used to build structures such as rammed earth construction. Adobe is soil moistened with water and sometimes mixed with chopped straw or other fibers added for strength. This mixture is placed into molds to create sundried blocks for building construction. Soil materials used for adobe construction consist of sand and clay. Ideally, the best adobe soil will have between 15% to 30% clay to bind the material together. Too much clay will cause the adobe to shrink and crack, while too little will cause the block to come apart. Sometimes, small amounts of cement or asphalt emulsion are added to the soil mixture to keep it intact during excessive wet weather. Detailed information on adobe construction is provided in Smith, 1982 and Masterson, 1975.

Rammed earth construction is similar to adobe construction with the difference that the soil mix in this type of construction is compacted within a formwork to create vertical walls. These forms are then removed leaving solid earth walls. The soil used is a processed engineered material consisting of screened soil. Small amounts of cement or asphalt emulsion is then added to this processed material. The mixture is placed in 8 inch layers within the formwork, and compacted to 5 inch thicknesses. Home building information using adobe and rammed earth construction can obtained from Hartworks, 2001 available at <http://www.greenhomebuilding.com>.

Reused soil, if used in adobe or rammed earth construction, needs to be processed. This processing involves screening, and often adding admixtures of straw, cement, asphalt, or other specified materials to fabricate the specified engineering properties.

Workers involved in the mixing, screening, and processing operations of the reused soils would have exposure to the reused soil. Another exposure to the reused soil for construction materials involves occupants of the adobe or rammed earth structure. However, this exposure would be much less than the workers who have close proximity when processing the reused soil.

5.6.1.3 Non-Radiologically Contaminated Soils

In order to better understand soil reuse, the researchers examined the construction related use of non-radiologically contaminated soil. A recent publication on the reuse and recycling of non-radiologically contaminated soils discusses the regulatory aspects, reuse and recycling technologies, laboratory and field test methods, and chemical characterization of the contaminated soil (Testa, 1997). The definition of contaminated soils varies from State to State, but in general, contaminated soil is defined as soil containing one or more contaminants from an intentional or unintentional spillage, leakage, emptying, emitting, or dumping of a hazardous substance or pollutant at a concentration that fails to satisfy any applicable standard. For example, in New Jersey, soils are considered hazardous when the following criteria are met (see <http://www.state.nj.US/dep/ dshw/rrtp/contsoil.htm>):

- Soil tests positive for characteristics of a hazardous waste as defined by Title 40, Part 261, Subpart C, paragraphs 21–24 of the *Code of Federal Regulations* (CFR).

- Soil contains a listed waste, as defined by 40 CFR Part 261, Subpart D, paragraphs 31–33.

- Soil is a mixture of a solid waste (non-hazardous) and one or more hazard wastes listed in 40 CFR Part 261, Subpart D, paragraphs 31–33.

Sources of non-radiologically contaminated soil include businesses, retail establishments, governments, and individuals. Quantities vary greatly depending on site-specific situations, from a few tons to hundreds of tons (See <http://www.state.nj.US/dep/dshw/rrtp/contsoil.htm>.)

The remedial action options for non-radiologically contaminated soils also vary somewhat from State to State. In New Jersey, for example remediation actions include onsite remediation, offsite management as a hazardous waste and, in some instances, recycling hazardous soils treated on site after receiving all applicable permits. Hazardous soils being moved off site for management must be properly manifested, and transported by licensed and insured haulers to a facility that is authorized to accept hazardous soils (<http://www.state.nj.US/dep/dshw/rrtp/contsoil.htm>). New Jersey's Department of Environmental Protection issued revised guidance on the remediation of contaminated soil, which includes soil reuse (New Jersey, 1998, available at <http://www.state.nj.us/dep/srp/regs/soilguide/sgd 0-xii.pdf>).

The disposal and reuse options for contaminated soils are specified in State regulations. For example, Massachusetts has requirements, standards, management practices, and approvals for the testing, tracking, transport, reuse, and disposal of contaminated soils at landfills (Massachusetts Department of Environmental Protection, 1997). There are limits on the contaminant levels for the reuse of soils at lined or unlined landfills. Contaminated soils that satisfy the criteria established by the State, can be reused as covers or pre-capping contour materials. The contaminants in the soils can include heavy metals, hydrocarbons, PCBs and other compounds. In cases where the criteria cannot be met and consistent with State laws, contaminated soils can be disposed at lined or unlined landfills depending on the nature and concentration of the contaminants in the soil. The listing of contaminants does not include radioactively contaminated soil and its disposal management options (Table 1 of Massachusetts Department of Environmental Protection, 1997, available at <http://www.state.ma.us/dep/bwp/files/comm9701 .pdf>).

There are many publications (e.g., Testa, 1997) and magazines (e.g., AEHS, 2001) available for reviewing case studies and guidance on soil remediation.

5.6.1.4 Radiologically Contaminated Soils

The literature search yielded very little information on the disposal management and reuse of radioactively contaminated soils. The only exception was extensive references in the literature relating to the reuse or disposal of uranium tailings contaminated materials (e.g., in tailings dams and other fill applications) or byproduct contaminated materials and slags produced from mineral extraction processes (other than uranium). These types of radioactively contaminated materials, and related reuse scenarios analyses have been extensively studied elsewhere.

The NRC staff is also aware of reuse of uranium

mill tailings, which is not considered to be soil for the purposes of this report. NRC regulations in 10 CFR Part 40 address the tailings or waste produced by extracting or concentrating uranium or thorium for its source material content. Therefore, uranium and thorium mill tailings and waste from uranium leaching (i.e., solution extraction) processes are regulated. Most uranium mill tailings that were inadvertently used in construction have been remediated and placed in engineered disposal cells (U.S. DOE, 1997). Although this information does not concern reused soil, it does provide a perspective on how free fill has been used.

5.6.2 Transportation of Reused Soils

The transport of soil materials from their source to their destination varies depending on the scale and nature of the task at hand and the equipment available. Large-scale projects could involve bulldozers, backhoes, graders, scrapers, and other equipment to excavate, spread, and mix the soils; front- end loaders, backhoes, and other equipment to collect and load the soil onto dump trucks, which in turn transport the materials to the construction site (in the case of uncontaminated or reuse soil materials) or disposal sites or landfills (in the case of contaminated soils). Smaller- scale projects typically rely on smaller, single-axle dump trucks, pickup trucks, or trailer assemblies attached to pickup trucks or similar sized vehicles.

The American Association of State Highway and Transportation Officials (AASHTO) in their publication "Standard Specifications for Highway Bridges," 16th edition, 1996, discusses the standard truck and lane loadings on the roadways of bridges in the United States. The load from a standard truck is designated as an H loading, which consists of a two-axle truck or a corresponding lane loading on a bridge. The specific loading for a given truck is designated as H followed by a number indicating the gross weight (in tons) of the standard truck. In this description, standard H loading is designated as an H15, which is a truck with a single rear-axle

with a gross rear-axle load of 24,000 pounds (12 tons) or a truck with two rear axles with a gross load of 16,000 pounds (8 tons) per axle, each spaced 4 feet apart. The weights to be assigned in computing the dead gross weight of the truck include 100 lbs/cu ft for loose soil materials.

Typical dump trucks transport 6 to 8 cubic yards (8 to10 tons or 7.2 to 9 metric tons) of soil and belong to the single rear axle class of truck loadings. Larger dump trucks can transport 8 to 15 cubic yards (10 to 20 tons or 9 to18 metric tons) of soil or more. Smaller trucks, generally used by smaller contractors, transport 2 to 5 cubic yards (3 to 4 tons or 2.7 to 3.6 metric tons) of soil materials, while pickup trucks, depending on their size, can transport about 0.75 to 1.5 cubic yards of soil materials (see <http://www.cars.com>).

5.6.3 International Studies

The NRC staff identified a relevant study from the United Kingdom's Department of the Environment on "Derivation of UK Unconditional Clearance Levels for Solid Radioactively Contaminated Materials" (Hill et al., 1999). The report discussed "recycling/re-use of soils," focusing on scenarios involving residential gardeners, and landscape workers in public gardens and parks. Specifically, the report identified these scenarios as leading to higher doses to individuals than scenarios dealing with "recycling" of large volumes of soils (such as commercial agricultural workers). When contacted via email during the current study, the report's principal author indicated that the parameters for these scenarios were estimated and there were no literature sources.

Recently, the European Commission developed guidance on general clearance levels for practices, and issued recommendations by a panel of experts which, however, did not directly address reused soil (European Commission, 2000).

The United Nations Scientific Committee on the Effects of Atomic Radiation (UNSCEAR) published a report on *Sources and Effects of*

Ionizing Radiation, which identified instances of uranium mill tailings used in construction, and estimated radon emanations rates of various material such as uranium mill tailings and soil (UNSCEAR, 1977).

A symposium paper (Reisenweaver and Linsley, 2000) summarized IAEA development of principles and criteria for regulating the release of material from regulatory control. The paper focused on the clearance process; however, it did not specifically address reused soil.

The NRC staff also identified a series of somewhat relevant reports from the international Biosphere Model Validations Study II (BIOMOVS), which was published as a special issue of the *Journal of Environmental Radioactivity*, 42(2–3), (in Davis, ed., 1999). These reports cite Chernobyl data to discuss radionuclide transport in soils, including radionuclide wash-off from contaminated watersheds.

5.6.4 National Radiological and Technical Guidance

The NRC staff identified a recent national document on "Recommended Screening Limits for Contaminated Surface Soil and Review of Factors Relevant to Site-Specific Studies: NCRP Report No. 129," issued in 1999 by the National Council on Radiation Protection and Measurements (NCRP). Although the publication discussed a wide variety of soil-related scenarios and provided parameter values, it did not address soil reuse. Similarly, the U.S. Environmental Protection Agency (EPA) issued two reports on soil screening guidance for radionuclides; "Soil Screening Guidance for Radionuclides: User's Guide" (EPA/540-R-00-007), and "Soil Screening Guidance for Radionuclides: Technical Background Document" (EPA/540-R-00-006). As with the NCRP report, these documents provided useful information on indigenous, native soils, but not on reused soil scenarios. The EPA also issued an "Environmental Factors Handbook" in 1997, which provided parameter values for soil ingestion and inhalation, along with plant uptake values. A recent publication on soils, the *Handbook of Soil Science* (Sumner, 2000), provided excellent information on soil properties, soil transport processes and soil water movement. Similarly, Brady and Weil, 2000 discusses the nature and properties of soil.

Table 5.16 Phase 2 NRC Staff Final Information Source Selections

Association for Environmental Health and Sciences (AEHS), *Contaminated Soil, Sediment, & Water*, Amherst, MA, 2001, (information available at <http://www.aehsmag.com/> as of January 27, 2002).

Chen, S.Y., N. Ranek, S. Kamboj, J. Hensley and A. Wallo, "Authorized Release of DOE's Non-Real Property: Process and Approach," *Health Physics*, 77(2) Supplement, pp. S40–S48, August 1999.

Davis, P. (ed.), "Special Issue: BIOMOVS II," *Journal of Environmental Radioactivity*, 42(2–3), Elsevier Press, New York, NY, 1999.

European Commission, Directorate-General for the Environment, "Radiation Protection 122: Practical Use of the Concepts of Clearance and Exemption—Part 1, Guidance on General Clearance Levels for Practices," Luxembourg, Luxembourg, 2000.

European Commission, Directorate-General for the Environment, "Radiation Protection 124: Radiological Considerations with Regard to the Remediation of Areas Affected by Lasting Radiation Exposure as a Result of a Past or Old Practice or Work Activity," Luxembourg, Luxembourg, 2001.

Gofman, J.W., *Radiation and Human Health*, Pantheon Books, New York, NY, 1983.

Table 5.16 Phase 2 NRC Staff Final Information Source Selections (continued)

Hill, M.D., M.C. Thorne, P. Williams, and P. Leyshon-Jones, "Derivation of UK Unconditional Clearance Levels for Solid Radioactively Contaminated Materials," DETR Report No. DETRA/RAS/98.004, U.K. Department of the Environment, W.S. Atkins & Electrowatt, UK, March 1999.

International Atomic Energy Agency (IAEA), "Principles for the Exemption of Radiation Sources and Practices from Regulatory Control," Safety Series No. 89, Vienna, Austria, 1988.

IAEA, "Application of Exemption Principles to the Recycle and Reuse of Materials from Nuclear Facilities," Safety Series No. 111-P-1.1, Vienna, Austria, 1992.

IAEA, "Clearance Levels for Radionuclides in Solid Materials, Application of Exemption Principles," IAEA - TECDOC 855, Vienna, Austria, 1996.

IAEA, "Technologies for Remediation of Radioactively Contaminated Sites," IAEA-TECDOC-1086, Vienna, Austria, June 1999.

Kennedy, W.E., Jr., and D.L. Strenge. "Residual Radioactive Contamination From Decommissioning: Technical Basis for Translating Contamination Levels to Annual Total Effective Dose Equivalent-Final Report," Volume 1, NUREG/CR-5512, U.S. Nuclear Regulatory Commission, Washington, DC, 1992.

Los Alamos National Laboratory, "MCNP 4A Monte Carlo N-Particle Transport System, RSIC Computer Code Collection (CCC) 200)," Los Alamos, NM, 1995.

Massachusetts Department of Environmental Protection, "Reuse and Disposal of Contaminated Soil at Massachusetts Landfills," Policy #COMM-97-001," Boston, MA, August 15, 1997, available at <http://www.state.ma.us/dep/bwp/files/comm9701.pdf>.

Meyer, P.D., and G.W. Gee, "Information on Hydrologic Conceptual Models, Parameters, Uncertainty Analysis, and Data Sources for Dose Assessments at Decommissioning Sites," NUREG/CR-6656, PNNL-13091, U.S. Nuclear Regulatory Commission, Washington, DC, December 1999.

National Council on Radiation Protection and Measurements, "Recommended Screening Limits for Contaminated Surface Soil and Review of Factors Relevant to Site-Specific Studies," Report No.129, Bethesda, MD, 1999.

National Radiological Protection Board (NRPB), "Radiological Protection Objectives for Land Contaminated with Radionuclides," Volume 9, No. 2, Chilton, Didcot, Oxon, United Kingdom, 1998.

New Jersey Department of Environmental Protection, Maps and Publications, "Guidance Document for the Remediation of Contaminated Soils," Trenton, NJ, 1998, available at <http://www.state.nj.us/dep/srp/regs/soilguide/sgd0-xii.pdf>.

Nicholson, T.J., and J. Parrott, "Proceedings of the Workshop on Review of Dose Modeling Methods for Demonstration of Compliance with the Radiological Criteria for License Termination," NUREG/CP-0163, U.S. Nuclear Regulatory Commission, Washington, DC, May 1998.

Table 5.16 Phase 2 NRC Staff Final Information Source Selections (continued)

Nisbet, A.F., and R.F. Woodman, "Soil-to-Plant Transfer Factors for Radiocesium and Radiostrontium in Agricultural Systems," *Health Physics*, 78(3):279–288, March 2000.

Oak Ridge National Laboratory, "Limiting Values of Radionuclide Intake and Air Concentration and Dose Conversion Factors for Inhalation, Submersion, and Ingestion," Federal Guidance Report No. 11, Oak Ridge, TN, 1988.

Pennsylvania Department of Environmental Protection, Bureau of Land Recycling and Waste Management, "Policy and Procedure Establishing Criteria for Use of Uncontaminated Soils, Rock, Stone, Unused Brick and Block, Concrete and Used Asphalt as Clean Fill," Doc. No. 258-2182-773, Harrisburg, PA, February 29, 1996.

Regens, J.L., et al., "Modeling Radiological Risks to Human Health from Contaminated Soils: Comparing MEPAS, MMSOILS, and RESRAD," *Human Ecological Risk Assessment,* 6(5):777–788, Washington, DC, October 2000.

Reisenweaver, D.W., and G. Linsley, "International Development of Principles and Criteria for Regulating the Release of Material from Regulatory Control," Proceedings of the 4th U.S. Department of Energy International Decommissioning Symposium, June 12–16, 2000, Knoxville, TN, 2001.

Renauld, P., J. Real, H. Maubert and S. Roussel-Debet, "Dynamic Modeling of the Cesium, Strontium and Ruthenium Transfer to Grass and Vegetables," *Health Physics*, 76(5):495–509, May 1999.

Sheppard, S.C., and W.G. Evenden, "Variations in Transfer Factors for Stochastic Models: Soil-To-Plant Transfer," *Health Physics*, 72(5):727-733, 1997.

Shlein, Bernard (editor). The Health Physics and Radiological Health Handbook, Silva Spring, MD: Scinta Inc., 1992.

Simon, S.L., "Soil Ingestion by Humans: A Review of History, Data, and Etiology with Application to Risk Assessment of Radioactively Contaminated Soil," *Health Physics*, 74(6):647–651, 1998.

Smith, J.T., D.R.P. Leonard, J. Hilton, and P.G. Appleby. Towards a Generalized Model for the Primary and Secondary Contamination of Lakes by Chernobyl-Derived Radiocesium," *Health Physics*, 72(6):880–891, June 1997.

Sumner, M.E., *Handbook of Soil Science*, CRC Press, Boca Raton, Florida, 1999.

Syracuse Research Corporation, User's Guide for the Integrated Exposure Uptake Biokinetic Model for Lead in Children (IIEUBK) Windows« version. Washington, DC: U.S. Environmental Protection Agency, 2001.

Testa, S.M., *The Reuse and Recycling of Contaminated Soils*, Lewis Publishers, Boca Raton, Florida, 1997.

United Nations Scientific Committee on the Effects of Atomic Radiation (UNSCEAR), *Sources and Effects of Ionizing Radiation*, New York, NY, 1977.

U.S. Department of Energy, Office of Environmental Management, "Uranium Mill Tailings Remedial Action Project: Fiscal Year 1997 Annual Report to Stakeholders," U.S. DOE, Washington, DC, December 31, 1997.

U.S. Department of the Interior, Bureau of Reclamation, *Earth Manual*, Third Edition, Washington, DC, 1990.

U.S. Department of Agriculture, Natural Resources Conservation Service, "1997 National Resources Inventory," Revised December 2000, available at <http://www.nhq.nrcs.usda.gov//land/lgif/m5058l.gif> as of July 2001.

U.S. Environmental Protection Agency, Radiation Protection Program, "Clean Materials Program," available at <http://www.epa.gov/radiation/cleanmetals/> as of January 27, 2002.

U.S. Environmental Protection Agency, *Environmental Factors Handbook*, Washington, DC, 1997.

U.S. Environmental Protection Agency, *Evaluation of the Potential for Recycling of Scrap Metals from Nuclear Facilities*, Washington, DC, 1997.

U.S. Environmental Protection Agency, "Soil Screening Guidance for Radionuclides: User's Guide," EPA/540-R-00-007, Washington, DC, October 2000.

U.S. Environmental Protection Agency, "Soil Screening Guidance for Radionuclides: Technical Background Document," EPA/540-R-00-006, Washington, DC, October 2000.

U.S. Nuclear Regulatory Commission, "Residual Radioactive Contamination from Decommissioning: Technical Basis for Translating Contamination Levels to Annual Total Effective Dose Equivalent," NUREG/CR-5512, Volume 1, PNL-7994, Washington, DC, October 1992.

U.S. Nuclear Regulatory Commission, "Residual Radioactive Contamination from Decommissioning: User's Manual DandD Version 2.1," NUREG/CR-5512, Volume 2, SAND2001-0822P, Washington, DC, April 2001.

U.S. Nuclear Regulatory Commission, "Residual Radioactive Contamination from Decommissioning: Parameter Analysis, Draft Report for Comment," NUREG/CR-5512, Vol.3., SAND99-2148, Washington, DC, October 1999.

U.S. Nuclear Regulatory Commission, "Summary and Characterization of Public Comments on the Control of Solid Materials," NUREG/CR-6682, Washington, DC, September 2000.

U.S. Nuclear Regulatory Commission, "Probabilistic Modules for the RESRAD and RESRAD-BUILD Computer Codes: User's Guide for RESRAD Version 6.0," NUREG/CR-6692, Washington, DC, 2000.

U.S. Nuclear Regulatory Commission, "Generic Environmental Impact Statement in Support of Rulemaking on Radiological Criteria for License Termination of NRC-Licensed Nuclear Facilities," NUREG-1495, Volumes 1–3, Washington, DC, July 1997.

U.S. Nuclear Regulatory Commission, "Decision Methods for Dose Assessments to Comply with Radiological Criteria for License Termination," NUREG-1549, Washington, DC, July 1998.

Table 5.16 Phase 2 NRC Staff Final Information Source Selections (continued)

U.S. Nuclear Regulatory Commission, "NRC Information on Control of Solid Materials," available at <http:www.nrc.gov/NMSS/IMNS/controlsolids.html> as of August 23, 2001.

Wisconsin Department of Natural Resources (DNR), "Application of Soil Performance Standards Guidance," PUB-RR-676, Memorandum from Mark F. Giesfeldt to DNR Bureau for Remediation and Redevelopment Staff, Statewide, Madison, WI, October 8, 2001, available at <http://www.dnr.state.wi.us/org/aw/rr/archives/pubs/RR676.pfd> as of November 27, 2001.

Yu, C., et al., "Development of Probabilistic RESRAD 6.0 and RESRAD-BUILD 3.0 Computer Codes," NUREG/CR-6697, U.S. Nuclear Regulatory Commission, Washington, DC, 2000.

Yu, C., et al., "Manual for Implementing Residual Radioactive Material Guidelines Using RESRAD, Version 5.0," Argonne National Laboratory, Argonne, IL, 1993.

Yu, C., et al., "Users Manual for RESRAD Version 6," Argonne National Laboratory, Argonne, Illinois, 2001.

6. QUALITY ASSURANCE/QUALITY CONTROL PLAN

6.1 Construction and Approval of the Plan

By virtue of the continuously expanding base of available resources, information retrieval is a more qualitative than quantitative process. For example, it has been estimated that as many as 14,000 technical reports are written each day in the United States. Another complicating factor to the reliable retrieval of information is the ephemeral nature of some sources, notably those on the Internet. As a result, the Quality Assurance/Quality Control (QA/QC) plan developed for this project emphasized study methodology.

All systems, including information retrieval systems, are constrained in three ways by time, quality, and cost. Each constraint is operative at all times, although their relative importance varies. Quality measures for information retrieval are founded on two components, precision and recall. Precision refers to the percentage of valid or highly significant citations as a function of the total number of citations retrieved (recalled). Recall is evaluated as the percentage of significant publications retrieved relative to the total number of significant publications available at the time the search was performed.

Quality measures for any project therefore depend upon the stated information needs of the client, balanced with the other constraints of time and cost. For the purposes of this report, the search quality performance for Phase 1 emphasized recall over precision. A reasonably extensive search was required to support the information needs of the NRC staff. The research plan was constructed in a way that minimized, to the extent possible, the negative impacts of a high-recall comprehensive survey. Phase 2 quality requirements emphasized precision. In addition, Phase 2 required significant investigator analysis, synthesis, and interpretation.

The development of a quality assurance or quality control plan was required by the terms of the Interagency Agreement (IAA). NAL drafted the plan during the initial phase of the IAA and submitted it to the NRC staff for their review and approval. Suggested changes were incorporated into the plan. The plan was written for the comprehensive search process used in Phase 1. Modifications to the original plan were made to reflect the procedures of Phase 2 .

6.2 The Quality Assurance/ Quality Control Plan

The project's final QA/QC Plan, as approved by the NRC staff, is summarized in Sections 6.2.1 and 6.2.2.

6.2.1 Procedures for Collaborative Review

Procedures for collaborative review of literature survey results, information sources, and retrieval strategies were established using guidelines set in 1996 by the Reference and User Services Association of the American Library Association (RUSA, 1996). NRC and NAL staffs reviewed the initial search strategies, concepts, definitions and descriptions, search terms, and database selections. Initial survey results were also reviewed in titles-only format. Suggested modifications were made and the results were incorporated into the Phase 1 draft letter report for external peer review.

6.2.2 QA/QC Audit

External peer review by non-NAL library and information science and soil science professionals was an important quality control measure. Reviewer recommended changes were incorporated as needed. Additional search results were presented to the NRC staff for their consideration.

The Phase 1 draft letter report was provided to the external reviewers for the QA/QC audit. The draft letter report included search strategies, data sources, and a reasonable sampling of the titles-only results. The reviewers were asked to evaluate the strategy for completeness. Specifically, the search terms were evaluated for any missing concepts, the strategies were analyzed for logic and, finally, the reviewers were asked to assess the retrieved results for inclusiveness of seminal works.

NAL staff met with NRC staff to consider the comments from both the external and NRC reviewers. Refinements were made to searches, as needed, to ensure maximum precision and completeness. NRC staff then selected specific titles for complete citation retrieval which are included in the Phase 1 final letter report.

Copies of all Phase 1 Dialog® session logs will be maintained by NAL until November 2004, 5 years from the completion of the Phase 1 final letter report. Session logs include information about costs, search terms and databases searched. Also, copies of the following Phase 1 reports will be held by NAL until November 2004: draft letter report, supplement to the draft letter report, and the final letter report. NAL will also maintain copies of the Phase 2 draft letter report and final NUREG-1725 until November 2004.

6.3 Phase 1 External Review

A key element in the QA/QC Plan was to engage external (i.e., non-NAL) experts to review research procedures. Two potential benefits expected through independent review were: (1) to alert NAL and NRC researchers to concepts missed in strategy development; and (2) to identify important information sources that may have been overlooked. These benefits were best achieved through careful selection of experts for the external review, as described in the next subsection.

6.3.1 Selection of the Phase 1 External

Reviewers

NAL and NRC staff jointly determined that external reviewers should come from three key disciplines, including (1) soil science, (2) civil engineering, and (3) information science. Soil science experts provide the best opportunity to identify new technical terms that could be added to the search strategies to enhance recall of relevant citations. The field of civil engineering (with its focus on construction) provides many important scenarios for the reuse of soils. Finally, professional information specialists are best suited to understand the complex syntax and logical search strategy construction, and are the most knowledgeable about available information sources.

Having established the expertise needed in the reviewers, the NAL staff sought capable experts. It should be noted that, in all cases, the reviews were voluntarily conducted without payment.

The following experts were selected by the NAL and approved by the NRC staff to serve as external reviewers:

Dr. Dewayne Mays
Head, USDA, NRCS, Soil Survey Laboratory

Dr. Mays has a Ph.D. in Soil Science from the University of Nebraska and is currently heading the National Soil Survey Laboratory for the USDA, Natural Resources Conservation Service.

Carol H. Reese
American Society of Civil Engineers

Ms. Reese has a Master of Library Sciences degree. She has developed and is maintaining databases for the American Society of Civil Engineering and is responsible for indexing the Society's publications. In addition, Ms. Reese has 16 years of reference research experience at a University. She serves as a board member of the Engineering Division of the Special Libraries

Association.

Carla Long Casler
Arid Lands Information Center
University of Arizona

Ms. Casler has a Master of Library Sciences degree. She has compiled information resources on "Soils of Arid Regions of the United States and Israel," "World Desertification Bibliography," and other soil-related projects. Ms. Casler has had professional involvement in both the United States Agricultural Information Network (USAIN) and the International Association of Agricultural Information Specialists (IAALD). She has 10 years of online bibliographic search experience in the Arid Lands Information Center. In addition, for 8 years Ms. Casler served as the North American Representative to CAB International (a key database resource used in the current study).

6.3.2 Summary of Phase 1 Reviewer Comments and Actions Taken

NAL requested that each reviewer consider four specific elements in conducting the review for the project:

- specific terms used in the strategies and to suggest additional terms

- search string logic or construction

- database selection

- recall of highly relevant publications.

With respect to the last point, NAL asked the reviewers to identify any highly relevant literature that was not found in the materials under review.

The reviewers were sent the draft letter report, which provided a copy of the comprehensive search strategy concept sets and sample title listings. These titles were felt to be an adequate sampling for review.

The external reviewers provided comments on search terms, strategy syntax, and information sources.

One reviewer suggested the additional search terms "removal" and "cost" for concept sets V3 and V1, respectively. These enhancements were made to the strategy statements. A syntax error was also noted and corrected in the strategy statement for concept set P3.

The reviewers recommended assessing additional information sources including dissertations and theses; publications by the USGS, STN International, and the Canadian Institute for Scientific and Technical Information (CISTI).

Dissertations and theses are indexed in Dialog® file 35, *Dissertation Abstracts Online*, which was searched in 15 of the final 42 searches. USGS publications are indexed in Dialog® file 89, GeoRef, which was used in 25 of the final 42 searches. Database and information access providers STN and CISTI were also considered. The reviewer selected list of CISTI journal titles were indexed in many of the Dialog® files previously searched. Further screening suggested that the depth of coverage in these resources was not sufficient to warrant further analysis at this time. (They remain potential new sources should additional research become necessary.)

Information sources recommended by one reviewer included Web sites related to Chernobyl and Three Mile Island. Information about these sites was relayed to the NRC staff for their consideration.

Appendix B contains additional details of the external review, comments, and actions taken.

6.4 Phase 2 Procedures

The Phase 2 study was distinctly different in nature than the Phase 1 study and required a variation to the Phase 1 QA/QC plan. In Phase 1 NAL researchers primarily used structured data resources, and they developed and executed detailed strategies after carefully reviewing search terms, logic, syntax, and data resource selections. The goal of the Phase 1 study was to identify citable published sources documenting practices and parameters for the reuse of soil in the United States. These sources were not found.

In contrast, the goal of the Phase 2 research was to develop a realistic basis for parameter estimation and validation. The process involved sifting facts, trends, and anecdotal information from a wide range of sources. In this case, there were no structured query sets accessible for external review of search terms, logic, syntax, and data resource selections.

As in Phase 1, NAL researchers conferred at timely intervals with the NRC staff to collaboratively review progress and to refine search and retrieval procedures. Because Phase 2 was the *first of its kind* synthesis, NAL researchers relied to the extent possible on independent verification of facts and inferences through multiple source confirmation and, where possible, direct contact with experts and business representatives via telephone or email interviews.

6.4.1 Selection of Phase 2 Technical Peer Reviewers

Emphases of the Phase 2 peer review were focused on the interpretations presented in the final report, rather than the information research procedures. As a result, Phase 2 reviewers were selected for their combined expertise in environmental sciences, civil engineering, and agronomy. Again it should be noted that in all cases the reviews were voluntarily conducted without payment.

The following experts were selected by NAL and approved by the NRC staff to serve as external reviewers:

Eric Koglin

Senior Environmental Scientist
National Exposure Research Laboratory
U.S. Environmental Protection Agency

Mr. Koglin received his B.S. in geology from Indiana State University, and his M.S. in hydrology from the University of Arizona. He has worked as a Superfund Remedial Project Manager. Currently, Mr. Koglin is specializing in the testing and evaluation of analytical methods to detect environmental contaminants in the field.

Robert M. Lacey
Assistant Technical Director, Military Land Management
U.S. Army Construction Engineering Research Laboratory
U.S. Army Corps of Engineers

Mr. Lacey received his B.A. in geography and his M.S. in city and regional planning from Southern Illinois University. Mr. Lacey's research has focused on installation natural resource management emphasizing land use planning. As Chair of the Tri-Service Conservation Pillar, he coordinated natural and cultural resources research conducted throughout the Department of Defense. Through these and other assignments, Mr. Lacey has represented a broad range of environmental research issues related to natural and cultural resources management on military installations.

Rufus L. Chaney
Research Agronomist
Animal Manure and By-Products Laboratory
USDA, Agricultural Research Service

Dr. Chaney received his B.S. in chemistry from Heidelberg College and his Ph.D. in biochemistry from Purdue University. Mineral metabolism in soybeans was the focus of Dr. Chaney's graduate studies. His postdoctoral work examined the mechanism of iron uptake in plants. Currently Dr. Chaney is conducting research on the fate, food-chain transfer, and potential effects of soil microelements. These studies include: plant uptake of metals; speciation of metals in plants and bioavailability to animals; development of crops for the remediation of contaminated soils; testing the bioavailability of lead and other metals in soils, biosolids, and composts; and methods to reduce food-chain transfers of toxic metals. Dr. Chaney has advised state and federal agencies regarding development of regulations to govern the use of biosolids. Specifically, Dr. Chaney is a technical consultant to the cooperative ISCORS project for sewage sludge analysis.

6.4.2 Summary of Reviewer Comments and Actions Taken

The reviewers were provided with a copy of the final draft of NUREG-1725 and asked to comment upon the study methodology, scope of coverage, and to recommend useful information sources not cited in the report under review.

All three of the reviewers acknowledged the problematic nature of assembling information in the material flow of reused soils in open commerce. The scarcity of available information and the collection of such information in integrated sources (e.g., publications, Internet web pages) made this enterprise challenging, and the reviewers expressed encouragement on the effort made to integrate such information.

The external reviewers provided comments in 3 general categories:

- editorial and technical corrections,

- process-related recommendations, and

- suggestions relevant to the reuse of soils, but outside the scope of the information search.

The NAL and NRC staff revised the report to correct typographic errors and other format-related problems in the text provided to the external reviewers.

The majority of the process-related comments were in the form of additional documents, which the reviewers recommended be included in the final publication of the information search results. However, a number were not included, primarily for scope reasons. The NRC is searching for information related to common practices in use of soils, which may have been removed from areas at or around nuclear facilities. Many of the suggested references relate to use of virgin (non-contaminated) soils, and the inclusion of these references would obscure the construction of likely scenarios for soils reuse. A number of the suggested citations do provide useful information on the parameterization of soils reuse and have been included in the citations listed in this publication.

The last category of comments made suggestions on how to model scenarios, how to select parameters, how to compute exposures and other relevant topics to the dose analysis of potential soils reuse scenarios. These references and suggestions did not address the question of material flow of reused soils, but rather the impacts from exposure to such soils after "delivery" to their final location. Although this is relevant to the NRC's overall development of technical bases for decision making, they were outside of the scope of this document. However, these references and the suggestions made by the reviewers were forwarded to those NRC staff offices responsible for the development of the technical basis supporting decision making in this area.

7. SUMMARY

This reports presents results of a 2-year investigation to compile information on current practices with regard to soil reuse. This information is intended to supplement the technical basis for formulating and characterizing scenarios related to exposure to residual radioactivity in reused soils. This process was both iterative and interactive, and involved timely review of both search strategies and results. Using this approach, the investigation team refined and improved the research effort; however, this process turned out to be complex for many reasons. For example, there is a paucity of reliable and documented information. Moreover, this report is a *first of its kind* effort to address the realistic material flow of reused soils.

Although this study revealed that no methodology currently exists for formulating realistic soil reuse exposure scenarios and for assigning the appropriate parameter values, it did yield some noteworthy observations:

- Large-volume transactions (such as 1,000 cubic yards or more) are generally distributed to construction projects (e.g., roads, parking lots, building pads).

- Large-scale greenhouse and landscape operations have moved away from using natural soils to specially formulated soilless growth media.

- Market leaders in packaged potting "soils" use mineral and organic matter in their formulations.

- Medium to small greenhouse and landscape operations continue to use natural soil mixtures.

- Small-volume transactions involving natural soils are primarily distributed to small businesses and private home owners.

- There still remains a demand for free or low-cost soil for diverse uses by the general public.

In addition, the study found the following trends in the greenhouse and landscape operations:

- Industry seeks inexpensive and readily accessible alternative materials for use in place of reused soil.

- Limited and variable fertility associated with natural soils encourages the use of specially formulated soilless growth media.

- The high cost of long-distance transportation and the low inherent economic value of the reused soil encourages use of locally available sources.

This study yielded the following benefits:

- This study supported and complimented the scoping scenario characterizations under evaluation in a parallel NRC effort for dose modeling analyses pertaining to soil reuse.

- This study refined parameter values to reflect realism in modeling exercises, and further improvements are expected.

- The study emphasized the significant uncertainty in material flows and soil reuse characterization (i.e., probabilistic distributions of uncertainty are needed).

The study also identified human factor sources of uncertainties:

- Material flow and exposure scenarios are a function of human behavior, which varies because of geographic and environmental aspects of the locations where the exposures occur.

- Anecdotal sources were useful because of the shortage of documented information.

This report compiles, for the first time, data and information sources for parameters specific to soil reuse. This report also provides information that is relevant for generic, as well as site-specific dose assessments. This investigation was coordinated with a companion effort by an ISCORS study that is directed toward the disposition of sewage sludge and dose modeling.

8. REFERENCES

Addison, K., and M. Hiraga, "Building a Square Food Garden," *Journey to Forever Web site*, available at <http://journeytoforever.org/garden_sqft.html> as of November 27, 2001.

American Association of State Highway and Transportation Officials (AASHTO), "Standard Specifications for Highway Bridges," 16th edition, 1996.

Baker, K.F., "The U.C. System for Producing Healthy Container-Grown Plants," California Agricultural Experiment Station Manual 23, 1957.

Biocycle, "Mulch and Topsoil Fit the Bill in Florida," *Biocycle* (Journal of Composting and Recycling), 41(2):71–72, February 2000.

Boodley J. and R. Sheldrake, Cornell Peat-Lite Mixes for Commercial Plant Growing, Cornell University Cooperative Extension, Ithaca, NY, 1982.

Boyles, P., "Dirty Business," *The Granite State Consumer*, University of New Hampshire, Durham, NH, July 19, 2000, available at <http://ceinfo.unh.edu/Common/Documents/gsc71900.htm> as of November 27, 2001.

Brady, N., and R. Weil, *Elements of the Nature and Properties of Soils*, Prentice-Hall, Inc., Upper Saddle River, NJ, 2000.

Butterfield, B.W., *National Gardening Survey*, National Gardening Association, 180 Flynn Ave., Burlington, VT 05401, 1997.

Canada, International Development Research Centre, "Promoting Urban Agriculture: A Strategy Framework for Planners in North America, Europe, and Asia," 1994, available at <http://www.idrc.ca/cfp/rep09_e.html> as of November 27, 2001.

Canada, Office of Urban Agriculture, "City Farmer's Urban Agriculture Survey Results," May 20, 2000, available at <http://www.cityfarmer.org/surveyresults.html#surveyresults> as of January 16, 2002.

Cars.com Web site <http://www.cars.com > as of November 28, 2001.

Chaney, R.L. and J.A. Ryan, "Heavy Metals and Toxic Organic Pollutants in MSW-Composts: Research Results on Phytoavailability, Bioavailability," pp. 451-506, *In* H.A.J. Hoitink and H.M. Keener (eds.), *Science and Engineering of Composting: Design, Environmental, Microbiological and Utilization Aspects*, Ohio State University, Columbus, OH, 1993.

Chaney, R.L. and J.A. Ryan, "Risk Based Standards for Arsenic, Lead and Cadmium in Urban Soils," (ISBN 3-926959-63-0) DECHEMA, Frankfurt, Germany, 130 pp., 1994.

Chaney, R.L., J.A. Ryan, U. Kukier, S.L. Brown, G. Siebielec, M. Malik and J.S. Angle, "Heavy Metal Aspects of Compost Use," pp. 323-359, *in* P.J. Stofella and B.A. Kahn (eds), *Compost Utilization in Horticultural Cropping Systems*, CRC Press, Boca Raton, FL, 2001.

Clausnitzer, H., and M. Singer, "Intensive Land Preparation Emits Respirable Dust," *California Agriculture*, 51(2):27–30, University of California, 1997.

Clausnitzer, H., and M. Singer, "Mineralogy of Agricultural Source Soil and Respirable Dust in California," *Journal of Environmental Quality*, 28(5):1619–1628, October 1999.

Cohen, J.T., B.D. Beck, T.S. Bowers, R.L. Bornschein and E.J. Calabrese, "An Arsenic Exposure Model: Probabilistic Validation using Empirical Data," *Human Ecol. Risk Assess.* 4:341–377, 1998.

Craul, P., "Compost in Urban Soil Design," *Biocycle*, 38(11):39-40, 42, November 1997.

Davis, P. (ed.), "Special Issue: BIOMOVS II," *Journal of Environmental Radioactivity*, 42(2-3), Elsevier Press, New York, NY, 1999.

Deese Landscaping, Hauling and Grading, "Price List for Fort Mill and Rock Hill (North Carolina) Areas," 1998, available at <http://web.cetlink.net/~dirt/> as of January 1998.

Ehret, D. Et al., "Clay Addition to Soilless Media Promotes Growth and Yield of Greenhouse Crops," *HortScience,* 33(1):67–70, 1998.

EPM Communications, Inc., "A Blooming Trend Most Americans Are Gardeners," *Research Alert*, March 6, 1998.

European Commission, Directorate-General for the Environment, "Radiation Protection 122: Practical Use of the Concepts of Clearance and Exemption—Part 1, Guidance on General Clearance Levels for Practices," Luxembourg, Luxembourg, 2000.

European Commission, Directorate-General for the Environment, "Radiation Protection 124: Radiological Considerations with Regard to the Remediation of Areas Affected by Lasting Radiation Exposure as a Result of a Past or Old Practice or Work Activity," Luxembourg, Luxembourg, 2001.

Furillo, A., "California Farm Labor by the Numbers," *Sacramento Bee*, May 20, 2001, available at <http://www.sacbee.com/static/archive/news/projects/workers/20010520_numbers.html> as of May 21, 2001.

Gouin, F., "Selecting Organic Soil Amendments for Landscapes," *Biocycle*, 38(12):62-63, December 1997.

Greer, L. and S. Diver, "Organic Greenhouse Vegetable Production: Horticulture Systems Guide," Appropriate Technology Transfer for Rural Areas (ATTRA), Fayetteville, AR, January 2000, available at <http://www.attra.org/attra-pub/ ghveg.html> as of January 27, 2002.

Hall, Charlie, "Economic impact of the green industry in Texas," The Texas Nursery & Landscape Association, 2001, available at <http://www.txnla.org/pdf_files/impact.pdf> as of January 27, 2002.

Handreck, K., *Growing Media for Ornamental Plants and Turf*, University of New South Wales Press, page 119, New South Wales, Australia, 1994.

Hartworks, Inc., "Building Today for Tomorrow," PO Box 632, Crestone, CO 81131, 2001, available at <http://www

Heimlich, R., and C. Barnard, "Agricultural Adaptation to Urbanization: Farm Types in Northeast Metropolitan Areas," *Northeastern Journal of Agricultural and Resources Economics*, 21(1):50–60, 1992.

Hill, M.D., M.C. Thorne, P. Williams, and P. Leyshon-Jones, "Derivation of UK Unconditional Clearance Levels for Solid Radioactively Contaminated Materials," DETR Report No. DETRA/RAS/98.004, U.K. Department of the Environment, W.S. Atkins & Electrowatt, United Kingdom, March 1999.

Housing Assistance Council. "Abundant Fields, Meager Shelter: Findings From a Survey of Farm worker Housing in the Eastern Migrant Stream," 2001, available at <http://www.ruralhome.org/pubs/farm worker/meager/toc.htm#table> as of November 27, 2001.

Howell, D., "Premium Soils Set to Outgrow Mix-it-Yourself Solutions, (Scotts Co.)," *DSN Retailing Today 40:2, 33,* January 22, 2001.

Huinink, J.T.M., "Soil Quality Requirements for Use in Urban Environments," *Soil and Tillage Research*, Proceedings of the World Congress of Soil Science on "State of the Art in Soil Physics and in Soil Technology of Anthrophic Soils," Montpellier, France, August 20–26, 1998, 47(1–2):157–162, 1998.

Irrigation & Green Industry Network–Landscaping Web site, <http://www.igin.com/Landscaping/> as of December 18, 2001.

Massachusetts Department of Environmental Protection, "Reuse and Disposal of Contaminated Soil at Massachusetts Landfills," Policy #COMM-97-001, Boston, MA, August 15, 1997, available at <http://www.state.ma.us/dep/bwp/files/comm9701.pdf> as of December 14, 2001.

Masterson, R., "Building with Adobe Brick," *Studio Potter, 4(2):54–58*, Daniel Clark Foundation, Goffstown, NH, 1975.

McCarty, L.B., "Basic Guidelines for Sod Production in Florida," Florida Cooperative Extension Service, Institute of Food and Agricultural Sciences, Bulletin 260, University of Florida, Gainesville, FL, Revised June 1994.

McCarty, B., G. Landry Jr., J. Higgins, and L. Miller, "Sod Production in the Southern United States," Circular 702, 43 pp.,Clemson University Cooperative Extension Service, Clemson, SC, 1999.

Mehta, K., et al., "Findings from the United States Agricultural Workers Survey (NAWS) 1997–1998," 2000, available at <http://www.dol.gov/asp/programs/agworker/report_8.pdf> as of November 27, 2001.

Monroe-Santos, S., "Recent National Survey Show Status of Community Gardens in U.S." *Commercial Greening Review*, 8:12,17, 1998.

Moore, M., W. Crosswhite, and J. Hostetler, "Agricultural Water Use in the United States, 1950–85," *National Water Summary 1987*, USGS Water Supply Paper 2350, USGS, Reston, VA, 1997.

Mulch and Soil Council, "Voluntary Uniform Product Guidelines for Horticultural Mulches, Growing Media and Landscape Soils," National Bark & Soil Producers Association Document V–6.3, Manassas, VA, 2001, available at <http://www.nbspa.org/industry/NB_Std_Consol_V62.pdf> as of January 27, 2002.

National Council on Radiation Protection and Measurements, "Recommended Screening Limits for Contaminated Surface Soil and Review of Factors Relevant to Site-Specific Studies," Report No.129, Bethesda, MD, 1999.

National Radiological Protection Board (NRPB), "Radiological Protection Objectives for Land Contaminated with Radionuclides," Volume 9, No. 2, Chilton, Didcot, Oxon, United Kingdom, 1998.

New Jersey Department of Environmental Protection, Maps & Publications, "1998 Revised Guidance Document for the Remediation of Contaminated Soils," Trenton, NJ, 1998, available at <http://www.state.nj.us/dep/srp/regs/soilguide/sgd0-xii.pdf> and specifically for soil reuse at <http://www.state.nj.us/dep/srp/regs/soilguide/sgd53-66.pdf>. As of November 28, 2001, the Web site available at <http://www.anjr.com/newsite/buyrecycled/buy_rec_fr.html> is replaced by <http://www.anjr.com/Resources/ANJRResources/RecycledProductsGuide/Grounds& Recreation.htm> which has been reorganized since the initial access on November 5, 2001.

Nilsson, Philip, *Labor Time Data Handbook*, Nilsson Associates, 374 Hart Street, Southington, CT 06489, 1996.

Nugent, R., "The Significance of Urban Agriculture," *Urban Agriculture Notes*, City Farmer— Canada's Office of Urban Agriculture, Ottawa, Canada, 1997, available at <http://www.cityfarmer.org/racheldraft.html> as of August 13, 2000.

Occupational Safety and Health Administration, "OSHA's Field Sanitation Standard," 1992, available at <http://www.osha-slc.gov/OshDoc/Fact_data/FSNO92-25.html> as of November 27, 2001.

Ohio Environmental Protection Agency, "Ohio's Materials Exchange TOPSOIL," available at <http://www.epa.state.oh.us/opp/recyc/avail6.html> as of January 23, 2001.

Online Computer Library Center, Inc., Office of Research, "Web Characterization Project: Statistics," 1999, available at <http://www.oclc.org/oclc/research/projects/webstats/statistics.htm> as of January 23, 2001).

Reference and User Services Association (RUSA), "Guidelines for Behavioral Performance of Reference and Information Professionals," Section 4, American Library Association, January 1996, available at <http://www.ala.org/rusa/stnd_behavior.html> as of January 18, 2002.

Rodgers, A., "Growth Opportunities in Soils and Amendments," *Lawn & Garden Marketing*, 28(7):36, August 1989.

Rutgers Cooperative Extension, "Preventing Heat Stress in Agriculture," available at <http://www.cdc.gov/niosh/nasd/docs2/nj00800.html> as of November 27, 2001.

Ryan, J.A. and R.L. Chaney, "Issues of Risk Assessment and its Utility in Development of Soil Standards: The 503 Methodology as an Example," pp. 393–414, *in* R. Prost (ed.), *Contaminated Soils: Proc. Third*

International Symposium on Biogeochemistry of Trace Elements, Paris, May 15–19, 1995, Colloque No. 85, INRA Editions, Paris, France, 1997.

Ryan, J.A. and R.L. Chaney, "Regulation of Municipal Sewage Sludge under the Clean Water Act Section 503: A Model for Exposure and Risk Assessment for MSW-Compost," pp. 422–450, *in* H.A.J. Hoitink, and H.M. Keener (eds.), Science and Engineering of Composting: Design, Environmental, Microbiological and Utilization Aspects, Ohio State University, Columbus, OH, 1993.

Sanders, D., "Vegetable Crop Irrigation," North Carolina Cooperative Extension Service, 1997, available at <http://www.ces.ncsu.edu/depts/hort/hil/hil-33-e.html> as of November 27, 2001.

Sheldrake, R., "Artificial Mix Substrates Commonly Used in U.S.A," *Acta Horticulturae,* 99:47–49, 1980.

Schultz Company, "Potting Soils," available at <http://www.schultz.com/potsoil.htm> as of January 27, 2002.

Scotts Company, "Scotts Company Product Guide Winter 2001–2002," available at <http://www.scottscompany.com/gardening/ProdGuideGarden.cfm> as of January 27, 2002.

SMG Investor Relations: Business Segment, "North American Consumer: Consumer Growing Media Group," available at <http://www.smgnyse.com/html/growingmedia.cfm> as of January 27, 2002.

Smith, E.W., "Large-Scale Adobe-Brick Manufacturing in New Mexico," Circular (182): 49–56, New Mexico Bureau of Mines and Mineral Resources, Socorro, NM, 1982.

Solley, W., R. Pierce, and H. Perlman, "Estimated Use of Water in the United States in 1995," Circular 1200, U.S. Geological Survey, Reston, VA, 1998, available at <http://water.usgs.gov/watuse/pdf1995/html/> as of November 27, 2001.

Sommers, P., and J. Smith, "Promoting Urban Agriculture: A Strategy Framework for Planners in North America, Europe, and Asia," *Cities Feeding People*, CFP Report Series Report 9, International Development Research Centre, Ottawa, Canada, 1994, available at <http://www.idrc.ca/cfp/rep09_e.html> as of January 30, 2202.

Standards Australia International Ltd., *Soils for Landscaping and Garden Use*, Doc. #AS 4419-1998, Sydney, Australia, 1998, only the scope and abstract are available at <http://www.standards.com.au/catalogue/script/Details.asp?DocN=stds000022925> as of January 25, 2002.

Stanek, E.J., E.J. Calabrese, and M. Zorn, "Soil Ingestion Distributions for Monte Carlo Risk Assessment in Children," *Human Ecol Risk Assess.* 7:357–368, 2001.

Stanley, Tracey, "Meta-Searching on the Web," *Ariadne,* Issue 14, 1998, available at <http://www.ariadne.ac.uk/issue14/search-engines/> as of January 23, 2000.

Sumner, M.E., *Handbook of Soil Science*, CRC Press, Boca Raton, Florida, 1999.

Testa, S.M., *The Reuse and Recycling of Contaminated Soils*, Lewis Publishers, Boca Raton, FL, 1997.

Three–Z–Inc., "Soil Blends," Valley View, OH, available at <http://www.three-z.com/> as of December 18, 2001.

Turfgrass Producers International, "Farmland and Urban Soil Conservation Resulting from Cultivated Turfgrass Sod," Turfgrass Producers International, Rolling Meadows, IL, Revised 1995.

United Nations Scientific Committee on the Effects of Atomic Radiation (UNSCEAR), *Sources and Effects of Ionizing Radiation*, New York, NY, 1977.

U.S. Bureau of Labor Statistics, "Grounds Maintenance Workers," *Occupational Outlook Handbook*, 1997, available at <http://www.bls.gov/oco/ocos172.htm> as of December 18, 2001.

U.S. Civil Air Patrol, "AETC Policy on Fluid Replacement During Training in Hot Environments," 2001, available at <http://www.capnhq.gov/nhq/cp/encampments/AETC.htm> as of November 27, 2001.

U.S. Department of Agriculture, Economic Research Service, "Farm Labor: Farm Labor Data," available at <http://www.ers.usda.gov/briefing/FarmLabor/farmlabor/> as of December 7, 2000.

U.S. Department of Agriculture, National Agricultural Statistics Service, "1998 Farm & Ranch Irrigation Survey, Census of Agriculture, Tables 22 & 29," (The 1997 Census is available at <http://www.nass.usda.gov/census/census97/fris/tbl22.pdf>).

U.S. Department of Agriculture, National Agricultural Statistics Service, Agricultural Statistics Board, "Farm Labor," 1995–2001, available at <http://usda.mannlib.cornell.edu/reports/nassr/other/pfl-bb/> as of November 27, 2001.

U.S. Department of Agriculture, National Agricultural Statistics Service, "Usual Planting and Harvesting Dates for U.S. Field Crops," 1997, available at <http://usda.mannlib.cornell.edu/reports/nassr/field/planting/> as of November 27, 2001.

U.S. Department of Agriculture, National Agricultural Statistics Service, Florida Department of Agriculture and Consumer Sciences, and the University of Florida Institute of Food and Agricultural Sciences, "Florida Agriculture Farm Labor Report," 2001.

U.S. Department of Agriculture, Natural Resources Conservation Service, "1997 National Resources Inventory," Revised December 2000, available at <http://www.nhq.nrcs.usda.gov//land/lgif/m5058l.gif> as of July 2001.

U.S. Department of Agriculture, Natural Resources Conservation Service, "Soil Taxonomy: A Basic System of Soil Classification for Making and Interpreting Soil Surveys," Agriculture Handbook #436, second edition, 1999.

U.S. Department of Agriculture, Soil Conservation Service, *Soil Taxonomy: A Basic System of Soil Classification for Making and Interpreting Soil Surveys*, Agriculture Handbook No. 436, Washington, DC, 1951. (More recent soil maps are available at <http://www.nhq.nrcs.usda.gov/land/meta/m4025.html>.)

U.S. Department of Energy, Office of Environmental Management, "Uranium Mill Tailings Remedial Action Project: Fiscal Year 1997 Annual Report to Stakeholders," Washington, DC, December 31, 1997.

U.S. Department of the Interior, Bureau of Land Management (BLM), "How to Obtain Sand, Gravel and other Mineral Materials from BLM Administered Federal Lands," BLM-WO-GI-92-002-4140, April 1992, available at <http://imcg.wr.usgs.gov/usbmak/sndgrvl.html> as of January 27, 2002.

U.S. Department of the Interior, Bureau of Reclamation, *Earth Manual*, Third Edition, Washington, DC, 1990.

U.S. Department of Labor, "The National Agricultural Workers Survey," available at <http://www.dol.gov/asp/programs/agworker/naws.htm> as of March 28, 2001.

U.S. Environmental Protection Agency, "Technical Support Document for the 40 CFR 503 Standards for the Use or Disposal of Sewage sludge," Washington, DC, 1992.

U.S. Environmental Protection Agency, "40 CFR Part 257 et al. Standards for the Use or Disposal of Sewage Sludge; Final Rules," *Federal Register* 58(32):9248-9415, 1993.

U.S. Environmental Protection Agency, Radiation Protection Program, "Clean Materials Program," available at <http://www.epa.gov/radiation/cleanmetals/> as of January 27, 2002.

U.S. Environmental Protection Agency, *Environmental Factors Handbook*, Washington, DC, 1997.

U.S. Environmental Protection Agency, *Evaluation of the Potential for Recycling of Scrap Metals from Nuclear Facilities*, Washington, DC, 1997.

U.S. Environmental Protection Agency, "Soil Screening Guidance for Radionuclides: User's Guide," EPA/540-R-00-007, Washington, DC, October 2000.

U.S. Environmental Protection Agency, "Soil Screening Guidance for Radionuclides: Technical Background Document," EPA/540-R-00-006, Washington, DC, October 2000.

U.S. Environmental Protection Agency, Office of Waste Water Management, "A Guide to the Biosolids Risk Assessments for the EPA Part 503 Rule," EPA/832–B–93–005, Washington, DC, 1995, available at <http://www.epa.gov/owm/bio/503rule/> as of January 30, 2002.

U.S. Environmental Protection Agency, Office of Waste Water Management, "A Plain English Guide to the EPA Biosolids Rule," EPA/832/R–93/003, Washington, DC, 1994, available at <http://www.epa.gov/owm/bio/503pe/> as of January 30, 2002.

U.S. Environmental Protection Agency, Office of Water, "Where Does My Drinking Water Come From?," 2001, available at <http://www.epa.gov/safewater/dwh/where.html> as of November 27, 2001.

U.S. Geological Survey, Mineral Resources Program, "Mineral Commodity Summaries," available at <http://minerals.usgs.gov/ minerals/pubs/mcs/> as of January 30, 2001.

U.S. Nuclear Regulatory Commission, "Residual Radioactive Contamination from Decommissioning: Technical Basis for Translating Contamination Levels to Annual Total Effective Dose Equivalent," NUREG/CR-5512, Volume 1, PNL-7994, Washington, DC, October 1992.

U.S. Nuclear Regulatory Commission, "Residual Radioactive Contamination from Decommissioning: User's Manual DandD Version 2.1," NUREG/CR-5512, Volume 2, SAND2001-0822P, Washington, DC, April 2001.

U.S. Nuclear Regulatory Commission, "Residual Radioactive Contamination from Decommissioning: Parameter Analysis, Draft Report for Comment," NUREG/CR-5512, Vol.3., SAND99-2148, Washington, DC, October 1999.

U.S. Nuclear Regulatory Commission, "Summary and Characterization of Public Comments on the Control of Solid Materials," NUREG/CR-6682, Washington, DC, September 2000.

U.S. Nuclear Regulatory Commission, "Probabilistic Modules for the RESRAD and RESRAD-BUILD Computer Codes: User's Guide for RESRAD Version 6.0," NUREG/CR-6692, Washington, DC, 2000.

U.S. Nuclear Regulatory Commission, "Generic Environmental Impact Statement in Support of Rulemaking on Radiological Criteria for License Termination of NRC-Licensed Nuclear Facilities," NUREG-1495, Volumes 1–3, U.S. Nuclear Regulatory Commission, Washington, DC, July 1997.

U.S. Nuclear Regulatory Commission, "Decision Methods for Dose Assessment to Comply with Radiological Criteria for License Termination," NUREG-1549, Washington, DC, July 1998.

U.S. Nuclear Regulatory Commission, "NRC Information on Control of Solid Materials," available at <http:www.nrc.gov/NMSS/IMNS/controlsolids.html> as of August 23, 2001.

University of California, Berkeley, "Battling Heat Stress," 1999, available at <http://are.berkeley.edu/heat/heat2/trial1.html#findings> as of November 27, 2001.

University of California, Berkeley, "Battling Heat Stress in Agriculture," available at <http://are.berkeley.edu/heat/> as of November 27, 2001.

University of Northumbria at Newcastle (UNN), "Web Search Service Features," 1999, available at <http://www.unn.ac.uk/features.htm> as of 23 January 2000.

Vogel, K. and D. Block, "An Eye for Quality, Bringing Composted Feedstocks to Market," *Biocycle*, 39(10):33-34, October 1998.

Warnicke, D., "Analyzing Greenhouse Growth Media by the Saturation Extraction Method," *HortScience*, 21(2):223–225, 1986.

Wessling, S., "The Right Truck for the Right Job," Irrigation and Green Industry Network Web site, Landscape, Irrigation & Maintenance Contractors, <http://www.igin.com/Landscaping/ index.html> as of December 18, 2001.

Wisconsin Department of Natural Resources (DNR), "Application of Soil Performance Standards Guidance," PUB-RR-676, Memorandum from Mark F. Giesfeldt to DNR Bureau for Remediation and Redevelopment Staff, Statewide, Madison, WI, October 8, 2001, available at <http://www.dnr.state.wi.us/org/aw/rr/archives/pubs/RR676.pfd> as of November 27, 2001.

Yu, C., et al., "Development of Probabilistic RESRAD 6.0 and RESRAD-BUILD 3.0 Computer Codes," NUREG/CR-6697, U.S. Nuclear Regulatory Commission, Washington, DC, 2000.

Yu, C., et al., "Manual for Implementing Residual Radioactive Material Guidelines Using RESRAD, Version 5.0," Argonne National Laboratory, Argonne, IL, 1993.

Yu, C., et al., "Users Manual for RESRAD Version 6," Argonne National Laboratory, Argonne, Illinois, 2001.

APPENDIX A. DISCUSSION OF PUBLIC COMMENTS RECEIVED ON DRAFT NUREG-1725

A.1 Release of NUREG-1725 for Public Comment

The U.S. Nuclear Regulatory Commission (NRC) is examining its approach for controlling release of solid material at NRC-licensed facilities. As part of developing technical information to support a Commission policy decision in this area, the NRC published a *Federal Register* notice (FRN), dated July 19, 2000 (65 FR 44843), requesting public comments on Draft NUREG-1725, "Human Interactions with Reused Soil: A Literature Search." This FRN indicated that the public comment period would last until September 18, 2000. The specific purpose of the draft NUREG-1725 was to review existing literature on potential ways in which soils are currently reused in commerce or by the general public. This information would then be used in assessing the potential radiation exposures that could result if soil is released from NRC-licensed facilities.

To solicit further comments for identifying information sources and technical bases for characterizing soil reuse scenarios, copies of the subject report were sent to land-grant universities and selected Federal agency libraries. In addition, a second FRN, issued on September 7, 2000 (65 FR 54326), extended the public comment period to November 17, 2000.

A.2 Public Comments Received

The NRC received 190 public comments in response to the subject FRNs. These comments can be categorized and analyzed as follows.

A.2.1 Comments Outside the Scope of NUREG-1725

The majority of the comments received did not address the specific purpose of the FRN, which (as described in Section A.1 above) was to solicit additional information from the public as to ways in which soil is currently reused.

Although these comments did not provide information within the scope of the FRN, they did indicate a strong objection to the release of any soils — contaminated or otherwise — from NRC-licensed facilities. The NRC will consider these comments in the context of its broader examination of the agency's policies on control of solid material. It is anticipated that this examination will take place following a study that is currently being conducted for the NRC by the National Academy of Sciences, which is scheduled to be completed early in 2002.

A.2.2 Comments Suggesting References for Soil Reuse Scenario Characterization

Some comments suggested references, which the investigators reviewed for relevancy to soil reuse scenario characterization:

- The NRC staff had previously identified one of these references in Phase 2 of the information search. This reference, National Council on Radiation Protection and Measurements Report No. 129, "Recommended Screening Limits for Contaminated Surface Soil and Review of Factors Relevant to Site-Specific Studies," is mentioned in Section 5.

- Some comments suggested broadening the scope of the information search to include such information as soil reuse in urban areas. This type of information is part of the scope of the literature already reviewed, particularly in the context of soil reuse in construction, agriculture, and recreation, and is mentioned in Section 5.

- The investigators determined that the other recommended references were not relevant to soil reuse scenario characterization because they deal with cancer and radioactivity, rather than the different ways that soil can be reused.

A.2.3 Comments Related to Analytical Methods

Some commenters suggested that the NRC should investigate references that provide information on radionuclide transport in the environment. However, because these comments did not relate to the ways in which soil is reused, they were considered to be outside the scope of this document. Nonetheless, these suggestions will be reflected in the analytical methods that are being developed under a separate contract to assess the impacts of soil reuse.

In addition, some comments suggested editorial corrections and improvements to the draft version of NUREG-1725. These comments are reflected in this final NUREG.

A.2.4 Other Comments

A few comments included suggestions for text improvements and the use of terminology. Specifically, a commenter questioned the use of terminology, such as "adequate," "realistic," and "soil reuse." The aim of the draft report was to try to envelop the possible uses that exist or may exist for soils transported from NRC-licensed facilities. The information gleaned from this exercise will assist in defining which scenarios are possible, which are unlikely, and which are not supported by actual practice. The report defines "reuse" in this regulatory context. The final report has improved the explanations of terminology unique to this area.

Finally, one comment supported the usefulness of the literature review for the NRC's decisionmaking process.

APPENDIX B. PHASE 1 SEARCH STRATEGIES, RESULTS, AND EXTERNAL REVIEW

B.1 Introduction

In Phase 1, the investigators conducted extensive literature searches to identify human interaction with soil. In doing so, the investigators focused on locating published, verifiable sources of information, which is primarily indexed in bibliographic citation or full-text databases. These databases can be accessed through commercial vendors through either online services or compact disc (CD) formats.

For Phase 1 of this study, the investigators primarily relied on online access to the bibliographic citation and full-text databases, and the Dialog® Corporation was the primary online access provider. In some instances, the investigators searched databases using the SilverPlatter® CD platform. Records accessed using this interface are equivalent to those available from the Dialog® system.

This appendix provides exact details of the methodology employed in Phase 1 of this study. Section B. 2 describes the detailed search strategies used to search the literature. Section B.3 summarizes the Phase 1 results. Section B.4 discusses the Phase 1 external review and Section B.5 identifies Internet resources of general interest.

B.2 Search Strategies

The searches were organized into three categories, including: general terms (labeled G), particular scenarios (labeled P), and quantitative or volumetric terms (labeled V). Sections B.2.1 through B.2.3 describe each of these three categories, followed by each of the related concept sets (each of which is identified by a label and a narrative purpose description). The actual strategy then appears on a separate line beginning with "S" (the search command for

Dialog®), followed by a listing of the selected databases searched. See Section B.2.4 for more information about Dialog® command syntax.

B.2.1 General Terms, Labeled G

For the general category, the investigators designed six concept sets for searches to discover and retrieve records of human actions or activities that involve soils and were not specifically identified by the NRC or NAL staffs. Search results could identify particular additional "scenario" names or terms or help to quantify particular soil contact parameters.

Concept Set G1

The G1 concept set retrieved many records describing soil use. These results were provided in five separate files because of the high number of records retrieved.

S (SOIL? ? OR DIRT)/TI,DE,ID,SH,CC (2N) (USE OR USERS OR USING OR USES OR USED OR USAGE? ? OR REUSE OR REUSING OR REUSED)

Dialog® business-related databases selected and searched, by file number[8]:
7,9,33,63,67,119,139,146,148,194,211,258,
262,318,323,474,475,483,484,492,494,495,
496,497,571,583,603,624,632,633,634,638,
639,640,642,649,701,702,704,705,706,707,
708,713,714,716,718,719,721,724,738,739,
741,743

Dialog® science-related databases selected and

---

[8]The Dialog® file numbers are listed in the order actually used in the search. This order determines which database file will be used to download records in the case of duplicate records.

searched, by file number:
10,6,50,60,8,15,16,18,20,35,47,49,58,
64,68,77,87,89,92,99,103,109,110,111,118,
143,144,238,257,266,292,319,335,479,484,
535,553,559,608,635,636,655,764

Concept Set G2

The G2 concept set retrieved records about material flow and soil. Material flow refers to the transfer or movement of a material or substance within a physical or commercial environment.

S SOIL? ?/TI,DE,ID,SH,CC AND MATERIAL? ?()FLOW?

Dialog® science databases selected and searched, by file number:
5,203,6,50,60,34,440,63,484,2,8,35,40,44,
71,89,94,103,108,117,118,144,156,162,
266,292,315,340,347,348,351,353,652,653,654

Concept Set G3

The G3 concept set retrieved records covering all forms of soil processing or processes, with the exception of soil formation.

S ((SOIL? ? OR DIRT)/TI,DE,ID,SH,CC (2N) PROCESS???) NOT (SOIL? ? (2W) FORM?)

Dialog® business-related databases selected and searched, by file number:
118,63,119,266,2,16,19,108,148,636,240,
484,7,109,67,624,323,621,813,111,583,18,553,19
4,262,633,649,516,635,15,238,47,
51,64,92,211,479,139,474,705,727,733,141

Dialog® science-related databases selected and searched, by file number:
50,10,6,89,8,65,292,103,68,110,76,58,
143,77,40,41,87,60,29,357,99

Concept Set G4

The G4 concept set searched for records related to direct human contact with soils.

S (SOIL? ? OR DIRT)/TI,DE,ID,SH,CC (S) (CONTACT? OR EXPOSURE OR HUMAN? ? OR PEOPLE? ? OR PERSON? ? OR WORKER? ? OR LABORER? ? OR WORKMEN OR WORKMAN)

Dialog® business-related databases selected and searched, by file number:
7,9,15,16,20,33,47,49,93,115,139,146,148,180,24
8,258,474,475,484,492,494,496,497,498,535,584,
603,608,630,631,632,633,634,636,638,641,642,65
5,701,702,704,705,707,708,713,714,721,723,732,
733,734,736,738,740,743,781

Selected SilverPlatter® CD science-related databases:
AGRICOLA, 1970–1999/06[9]; CAB 1972–1999/04; NTIS, 1983–1999 1–18

Concept Set G5

The G5 concept set searched for records related to temporary soil storage (e.g., surcharge piles) or long-term warehousing of stored soil.

S (SOIL? ? OR DIRT)/TI,DE,ID,SH,CC (2N) (STORAGE OR STORED OR STORING OR DISPOS??? OR SURCHARGE()PILE? ?)

Dialog® business-related databases selected and searched, by file number:
194,118,315,63,60,266,119,109,98,323,
148,636,108,474,484,16,99,553,262,87,92,603,11
1,559,660,655,317,195,49,238,335,
479,527,635,492,634,707,737,319

Dialog® science-related databases selected and searched, by file number:
6,10,50,8,89,103,68,292,58,143,29,41,
2,35,96

---

[9]The SilverPlatter® CD-based databases are followed by date ranges of subject coverage; the two-digit number following the slash represents the release month for the CD.

Concept Set G6

The G6 concept set retrieved records from specific U.S. Government agencies with missions and responsibilities related to regulating radionuclides and describing soil.

S (SOIL? ? OR DIRT)/TI,DE,ID,SH,CC (F) (AEC or DOE (S) (FOCUS () GROUP or STABIL?) OR ERDA or NRC)

Dialog® science-related databases selected and searched, by file number:
10,6,203,5,50,65,2,73,76,89,123,108,109,
117,144,148,155,156,241,266,285,292,440,624,63
6,655,660

B.2.2 Particular Scenario Terms, Labeled P

For the particular scenarios, the investigators designed concept sets for 11 searches to retrieve items pertaining to specifically identified types of human-soil interactions.

Concept Set P1

The P1 concept set retrieved records on soil uses for golf courses and for sod farming and sod roof construction.

S SOIL? ?/TI,DE,ID,SH,CC (S) (GOLF()COURSE? ? OR (SOD OR TURF)()(FARM? ? OR ROOF? ?))

Dialog® databases selected and searched, by file number:
10,50,76,5,203,8,35,41,60,65,71,77,16,18,
19,47,103,143,144,266,286,292,479,516,
555,630,631,632,633,634,641,707,708,713,714,71
6,723,725,733,742,777,781,34,440

Concept Set P2

The P2 concept set is composed of two subsets combined to retrieve records describing techniques used in the cleanup of contaminated soils. The combined Boolean operation is displayed as:

S SOIL? ?/TI,DE,ID,SH,CC (3N)(REMEDIAT? OR RECLAM? OR RECLAIM??? OR WASH??? OR CLEAN??? OR PROCESS??? OR RECYCL??? OR STABILIZ?)
  -and-
S (METHOD? OR TECHNIQUE? ? OR MECHANISM? ? OR PROCEDUR?? OR OPTION?? OR PLAN???? OR ACTIV????? OR ACTION??)/TI,DE,ID,SH,CC

Dialog® science-related databases selected and searched, by file number:
10,6,50,76,203,8,15,16,18,19,35,40,41,58,
60,63,65,68,77,87,89,92,98,99,103,110,
111,117,118,144,148,238,266,285,292,315,317,48
4,527,535,553,559,621,624,636,660,764,766,813,
7,194,262,649

Concept Set P3

The P3 concept set is composed of two subsets combined to retrieve records describing dust from soil. The combined Boolean operation is displayed as:

S DUST? ?/TI,DE,ID,SH,CC (3N)(LOAD? OR LEVEL? ? OR VOLUME? ? OR QUANTIT? OR AMOUNT??? OR HAZARD? OR LOSS OR LOSSES OR DAMAG? OR TRANSFER? OR CONTAMINAT?) NOT (DUST (2N) (HOUSE? ? OR MITE? ?))
  -and-
S (SOIL? ? OR DIRT OR EARTH? ?) /TI,DE,ID,SH,CC

Dialog® databases selected and searched, by file number:
10,50,6,5,103,40,89,110,41,73,144,8,2,76,
337,117,155,68,474,65,109,63,655,108,
119,315,334,7,240,16,323,60,77,161,9,19,
31,99,317,535,636,111,118,262,475,747

Concept Set P4

The P4 concept set retrieved records covering the

use of soil in construction.

S (EARTHMOV??? OR EARTH()MOVING OR
RAMMED()EARTH OR (BACKFILL??? OR
FILL()DIRT OR (BACK OR CLEAN OR
CONSTRUCTION OR RESIDENTIAL ()
FILL))(F)(SOIL? ? OR DIRT OR
EARTH??)/TI,DE,ID,SH,CC

Dialog® business-related databases selected and
searched, by file number:
7,9,49,146,269,474,475,478,483,484,492,
494,495,527,535,553,555,559,570,603,608,621,62
4,632,634,563,633,636,638,639,640,641,642,649,
655,660,704,707,708,712,713,714,718,723,738,74
3,781,813

Dialog® science-related databases selected and
searched, by file number:
10,6,50,203,8,15,16,18,20,35,41,47,58,63,
64,180,194,195,196,257,285,635,636,14,
19,68,77,87,89,98,99,156,161,292,103,109,117,11
8,144,110

Concept Set P5-6

The P5-6 concept set retrieved records covering
soil use in walls, berms, dams, etc.

S (SOIL? ? OR EARTH?? OR
DIRT)/TI,DE,ID,SH,CC (2N) (REINFORC? OR
EMBANKMENT? ? OR DAM? ? OR LEVEE? ?
OR DIKE? ? OR BERM?? OR WALL?? OR
ADMIXTURE? ?)

Dialog® databases selected and searched, by file
number:
63,8,65,89,118,144,10,6,50,203,103,15,33,34,35,5
0,67,119,194,262,248,535,559,624,670,765,2,19,3
1,35,47,40,44,58,68,41,77,
87,92,96,98,89,110,117,430,238,99,240,
266,292,293,440

Concept Set P7

The P7 concept set included terms to retrieve
records describing adobe building materials and
construction.

S (ADOBE/TI,DE,ID,SH,CC NOT (SOFTWARE

OR PROGRAM??? OR COMPUT??? OR
DESKTOP? ? OR ILLUSTRATOR OR
PHOTOSHOP OR PRINTSHOP OR
ACROBAT))(F) (SOIL? ? OR DIRT OR
MATERIAL? ? OR SOURCE? ? OR MAKING)

Dialog® databases selected and searched, by file
number:
10,6,203,5,89,531,118,103,47,65,68,2,
148,634,35,111,475,16,87,99,292,492,
603,9,132,262,498,630,713,716,719,732

Concept Set P8

The P8 concept set retrieved records covering
pottery production or potter's clay.

S ((POTTING OR POTTERY OR
POTTERS)()CLAY? ?)/TI,DE,ID,SH,CC OR
((POTTING OR POTTERY OR POTTERS)()
CLAY? ?)(F)(SOURCE? ? OR SITE? ? OR
PRODUC??? OR SUPPL???? OR
MANUFACTUR?)

Dialog® databases selected and searched, by file
number:
5,6,89,47,2,15,117,20,63,571,58,8,103,109,
118,146,148,535,583,483,704,708,716,717,
719,724,608,632,519,633,634,638,641,642,
718,781,736,702,703,706,725,734,492,
494,737,248,335,624,723,733,740,741,743

Concept Set P9

The P9 concept set was designed to retrieve
records covering detrital materials; however, no
relevant material was found.

S (DETRIT?? AND (SOIL OR DIRT OR
EARTH??))/TI,DE,ID,SH,CC

Concept Set P10

The P10 concept set retrieved records discussing
soil erosion rates. Additional related documents
from the USDA, NRCS Web-site were also
retrieved.

S SOIL()EROSION/TI,DE,ID,SH,CC (F) RATE

Selected SilverPlatter® CD database: AGRICOLA (1970–1999/06)

Concept Set P11

The P11 concept set retrieved records on bulk or packaged soil mixes, including potting soil.

S SOIL? ?/TI,DE,ID,SH,CC (2N) (BULK?? OR PACKAG??? OR BAGGED OR BAGGING OR MIX??? OR POTTING)

Dialog® business-related databases selected and searched, by file number:
18,67,116,119,141,194,358,474,475,478,
484,492,494,495,527,531,559,603,632,633,634,63
9,640,701,702,703,704,705,706,707,708,712,713,
714,715,718,720,724,731,735,736,781,813,861,73
3,9,15,647,285,319,
479,535,553,621,624,635,766,7

Dialog® science-related databases selected and searched, by file number:
10,50,5,6,60,8,63,68,89,103,117,143,
144,285,292,516,515

Concept Set P12

The P12 concept set retrieved records with topsoil as a subject term, excluding records retrieved from other searches using soil and dirt terms.

S TOPSOIL? ? NOT (SOIL? ? OR DIRT))/TI,DE,ID,SH,CCV)

Dialog® databases selected and searched, by file number:
50,6,76,34,40,41,47,2,9,44,58,68,77,89,92,99,103,
110,117,118,143,144,148,180,238,
285,479,484,516,535,571,608,624,635,636,637,66
5

B.2.3 Volumetric Terms, Labeled V

For the quantitative or volumetric category, the investigators designed concept sets for three searches. Using volume, quantity, or economic terms, these searches retrieved records to quantify, specify, or delineate the extent of human contact with soils.

Concept Set V1

The V1 concept set is composed of three subsets combined to retrieve records covering the economic literature for soils, while excluding the economic discussions regarding soil erosion and conservation, soil fertility, pesticides, soil surveys, etc. The combined Boolean operation is displayed as:

S (SOIL? ? OR DIRT)/TI,DE,ID,SH,CC (S)(ECONOM? OR DOLLAR? ? OR PRICE? ? OR PRICING OR PAYMENT? ? OR EXPENS? OR CASH OR VALU????? OR BUSINESS?? OR RETAIL?)
  -or-
S (SOIL? ? OR DIRT)/TI,DE,ID,SH,CC (S) (WHOLESALE? OR PROFIT? OR COST??? OR INDUSTR??? OR COMMERC??? OR BUSINESS?? OR INVEST? OR MARKET??? OR SALE? ? OR PURCHAS??? OR DOLLAR? ?)
  -not-
S (EROSION OR EROD? OR FERTIL? OR LOSS OR LOSSES OR POLLUT? OR RECLAM? OR RECLAIM? OR SAMPL? OR INVESTIGAT? OR CONSERV? OR SOIL()SURVEY? ?)

Dialog® business-related databases selected and searched, by file number:
7,63,139,47,474,111,484,20,16,635,713,
483,603,553,110,99,636,2,29,660,49,705,
642,18,87,475,634,98,621,258,631,718,
624,119,632,633,19,92,103,559,148,
531,285,194,266,109,9,479,47,535,765

Dialog® science-related databases selected and searched, by file number:
10,50,6,89,103,60,292,68,65,8,143,266,35,58,40,7
7,109,118

Concept Set V2

The V2 concept retrieved records with specified numeric data for soils, excluding topics related to erosion, pesticides, and fertility.

S (SOIL? ? OR DIRT)/TI,DE,ID,SH,CC (2N)

(QUANTIT? OR STATISTIC? OR AMOUNT? ?
OR WEIGH? OR VOLUME? ?) NOT (EROSION
OR EROSIV??? OR EROD??? OR LOS??? OR
WEIGHTED OR FERTILI? OR YIELD??? OR
PESTICIDE? ? OR HERBICIDE? ?)

Dialog® business-related databases selected and
searched, by file number:
763,240,118,119,109,108,2,92,266,716,
194,99,484,7,474,609,708,357,148,248,
315,317,636,483,738,9,16,111,553,559,
475,494,633

Dialog® science-related databases selected and
searched, by file number:
10,50,6,103,68,58,292,143,41,110,40,60,
98,29

Concept Set V3

The V3 concept set is composed of two subsets
to retreive records covering soil movement,
shipment, or transportation, but excluding
movement of fertilizer elements, pesticides, or
other chemicals applied to soils. The combined
Boolearn operation is displayed as:

S (SOIL? ? OR DIRT)/TI,DE,ID,SH,CC (F)
(TRUCK? OR SHIP? OR TRANSPORT? OR
HAUL? OR BARG? OR TRAIN? OR RAIL? OR
CONVEY??? OR   REMOV? OR
RELOCAT? OR REPLAC? OR PLACE? ? OR
PLACEMENT)
-not-
S (FERTIL? OR CHEMICAL? ? OR
INSECTICIDE? ? OR SEED? ? OR NUTRIENT? ?
OR PESTICIDE? ? OR HERBICIDE? ?)

Dialog® databases selected and searched, by file
number: 10,6,50,2,58,67,68,89,103,109,143,292,9,
15,16,18,19,20,33,47,49,63,64,92,98,99,
111,118,119,180,211,238,240,241,245,248,266,26
9,479,516,527,535,553,559,570,571,608,621,624,
635,636,637,660,813,7,474,
475,258,262

B.2.4 Explanation of Dialog® Search Command Syntax

The following sections describe the Dialog®
search command syntax used in this study, as
reported in Sections B.2.1 through B.2.3, above.

The Search Command

The Dialog® information system can perform
many types of operations. The operation to
search files for records is initiated using the
search command. The syntax for the search
command is an "S", placed at the beginning of
each statement. The statement identifies terms
and operations used in the search.

Truncation Command

Word truncation is a method used to capture
spelling variations, such as suffixes of -ed or -ly
added to the root word. The Dialog® system
truncation command, "?", can be used in the
following ways:

- A single "?" will retrieve all records containing
  the root word.   This use of the command
  allows an indeterminate number of characters
  to follow the root word.

- A double "?" ("? ?") can be used to limit
  spelling variations to no more than one
  character after the root word.

- Additional "?" commands, such as "???" or
  "????", can limit the ending length of the root
  word for any number of characters up to one
  less than the number of "?" commands
  shown.

Boolean Operators

Boolean operators "and", "or", or "not" specify
whether terms occurring on either side of the
operator must be, may be, or cannot be contained
within a record, respectively.

Suffix-Coded Field Tags

Dialog® databases are generally structured into
specific fields identified with tags. It is possible to
use the database field tags to limit searching to
specific fields. This type of limitation can generally
improve the relevance of the search findings.

The Dialog® syntax for using suffix-coded field tags is a "/" followed by the field name abbreviation [/TI, DE, ID, SH, CC]. Terms appearing immediately before the "/" must be present within a specified field. The fields used here include TI = titles, DE = descriptors (subject terms), ID = identifiers, SH = subject headings and CC = category codes.

Proximity Operators

Proximity operators [(F), (S), (3N), or ()] indicate the allowed location of terms within a record. (F) requires that terms on either side of the operator must be in the same field; (S) requires that terms must be in the same subfield (i.e. in the same phrase or "sentence"); (nN) requires that terms on either side of the operator must be separated by not more than "n" terms, where "n" is a number, in any order; the "()" operator requires that terms on either side of the operator must be both adjacent and in the specified order.

Parenthetical Grouping

Parentheses group terms together for processing. Such grouping is used for Boolean operations (and, or, not), or to apply field search limits (/TI, DE, ID, SH, CC) or proximity operators [(F), (S), (3N), etc.] to all terms within a parenthetical group.

Command operations are performed first within parenthetical groupings before any other operations are processed. This command syntax is analogous to the precedence of operations seen in mathematical equations.

Field Limitations

Proximity operators that search for terms within a given field or subfield will by default limit other linked but unlimited terms to the same field or subfield.

For example, searching a given set of terms that have been limited to fields, as in /TI,DE,ID,SH,CC, when linked to another term or parenthetical group using (F) or (S) requires all terms in the second group must also occur in one of the specified fields (/TI,DE,ID,SH,CC), by virtue of the (F) or (S) requirements.

B.3 Study Results

From more than two million database records initially found in surveys of Dialog® databases, the NAL investigators presented approximately 78,000 items to the NRC from results of the searches outlined in the previous section. The majority of these items were titles that were provided to the NRC staff in electronic format.

An inventory of the complete count of items retrieved by the completed searches is shown in Table B.1. Because of the large amount of data retrieved, processing limits of the Dialog® system did not allow all of the results from many of the individual concept sets to be included into single files. Therefore, many concept set results were split into two or more files as seen in Table B.1. Specific details on these processing limits and the techniques that were used to separate the data into multiple files are presented at the end of the table.

Table B.1 Concept Set Findings

Concept Set	Included Files	File Size in Bytes	Count
G1 *Soil use*	G1BIZ.TTL G1SCI1.TTL G1SCI2.TTL G1SCI3.TTL G1SCI4.TTL	146,325 652,687 668,994 819,476 704,757 (2,992,239)	696 titles 2,797 titles 2,928 titles 3,624 titles 2,379 titles (12,424 titles)
G2 *Soil material flow*	G2SCI.TTL	116,521 (116,521)	502 titles (502 titles)
G3 *Soil process (not soil forming)*	G3BIZ.TTL G3SCI.TTL	114,157 32,221 (146,378)	574 titles 145 titles (719 titles)
G4 *Human contact with soil*	G4AGR.TTL G4AGR.TXT G4BIZ.TTL G4CAB.TTL G4CAB.TXT G4NTIS.TTL G4NTIS.TXT	97,511 43,427 189,395 91,292 85,997 133,183 184,476 (825,283)	571 titles (CD) 49 selected records 442 titles (CD) 559 titles (CD) 19 selected records 724 titles (CD) 40 selected records (2,296 titles) (108 records)
G5 *Storing soil*	G5BIZ.TTL G5SCI.TTL	119,052 963,594 (1,082,646)	587 titles 4,379 titles (4,966 titles)
G6 *Publication on soil from applicable Federal Agencies*	G6.TTL	64,237 (64,237)	295 titles (295 titles)
P1 *Golf courses and sods*	P1.TTL	30,172 (30,172)	150 titles (150 titles)
P2 *Reclamation methods*	P2ALL1.TTL P2ALL2.TTL	534,222 480,764 (1,014,986)	2,747 titles 2,396 titles (5,143 titles)
P3 *Soil dust*	P3.TTL	118,490 (118,490)	516 titles (516 titles)

Table B.1 Concept Set Findings (continued)

Concept Set	Included Files	File Size in Bytes	Count
P4 *Earthmoving and soil use in construction fill and rammed earth*	P4BIZ.TTL P4SCI.TTL	244,778 480,764 (697,177)	1,158 titles 2,230 titles (3,388 titles)
P5-6 [a] *Soil in walls, dams, berms, and dikes*	P56BIG6.TTL P56BIG62.TTL P56NTIS.TTL P56OTHER.TTL	916,776 1,200,346 444,637 562,442 (3,124,201)	4,038 titles 5,485 titles 1,787 titles 2,889 titles (14,199 titles)
P7 *Adobe*	P7.TTL	34,755 (34,755)	177 titles (177 titles)
P8 *Pottery production and potting clay*	P8.TTL	27,035 (27,035)	152 titles (152 titles)
P10 *Soil erosion rates*	P10AGRIC.TXT P10WEB.TXT	19,620 22,616 (42,236)	39 titles Web Resources [b] (39 records)
P11 *Potting soil and bagged or bulk soil*	P11BIZ.TTL P11SCI.TTL	82,003 587,434 (669, 537)	433 titles 2,821titles (3,254 titles)
P12 *Topsoil*	P12ALL.TTL	134,414 (134,414)	278 titles (278 titles)
V1 *Soil economics, business activities*	V1ABIZ.TTL V1ASCI.TTL V1BBIZ.TTL V1BSCI.TTL	286,444 961.951 826,537 220,487 (2,295,349)	953 titles 4,034 titles 3,384 titles 992 titles (9,363 titles)
V2 *Statistical and numeric data for soils*	V2BIZ.TTL V2SCI.TTL V2SCI89.TTL	132,438 513,626 8,589 (654,653)	641 titles 2,069 titles 35 (sample titles) (2,745 titles)
V3 *Soil transportation*	V3BIZ.TTL V3SCIPRT.TXT V3SCIB1.TXT V3SCIB2.TXT V3SCIB3.TXT V3SCIB4.TXT	788,143 49,696 701,727 791,884 687,442 871,151 (837,839)	3,554 titles 205 (sample titles) [c] 3,040 titles 3,372 titles 3,074 titles 4,123 titles (17,163 titles)

Table B.1 Concept Set Findings (continued)

Concept Set	Included Files	File Size in Bytes	Count
Totals *19 concept set results*	42 search files	14,908,148 bytes	77,730 titles, plus 147 complete citations

[a] P5-6 is represented by P56 in the electronic file name for the consistency in file nomenclature.
[b] P10 results included USDA, NRCS Web pages and links.
[c] This file only included a sampling of titles. Complete listings were provided to the staff of the NRC at a later date in files V3SCIB(1-4).TXT. The 205 titles are encompassed within the complete files and are not counted in the total.

For the purposes of this project, two Dialog® system limitations impacted the processing of the searches. Specifically, these limitations were (1) the total number of files that could simultaneously be searched, and (2) the total number of items that could be processed to remove duplicate records.

The Dialog® system limits multiple database searching to a maximum of 60 simultaneous files; however, some search concepts required exploration of more than 60 files. In those cases, the same search strategy was run several times against different groups of databases until all selected databases had been searched. Often, databases were grouped into science-focused "SCI" or business-focused "BIZ" categories.

Overlap of literature coverage exists between databases where the same journals are indexed. Dialog® can process the removal of duplicate records from multiple database searches up to a maximum of 5,000 items; however, this limit was often exceeded. In those cases, the records were separated into groups of less than 5,000. In most cases, appropriate groupings were made using publication dates as group delimiters.

When particular searches included the use of SilverPlatter® CD versions of specific databases, results files from each database were kept separate. This was done because duplicates among multiple databases could not be removed by processing commands within the SilverPlatter® system. For these various reasons, many results groups listed in Table B.1 include more than one file.

In addition to the comprehensive database output described in this appendix, research results came from review and selection of specific resources from the Internet, WorldCat® (OCLC's comprehensive national multi-library database), Statistical Masterfile on CD, DTIC databases, and the Thomas Register of American ManufacturersSM database. These results and processes are described in Section 4. A specific list of selected Internet resource URLs is shown in Section B.5.

B.3.1 Dialog® Database File List

With more than 500 databases, Dialog® is one of the most comprehensive information resources available today. The Dialog® system contains more than 330 million articles, abstracts, and citations with information covering a broad spectrum of topics.

This valuable resource was extensively searched for literature describing how humans come into contact with soil, how soil is used (or reused), how contaminated soils are cleaned or reclaimed, and what models are used to calculate potential exposures.

The databases searched were carefully selected using the following criteria:

- journals indexed,
- focus or scope of the database, and

- date ranges of database material.

The validity of the database selections was further verified by sampling the search output for relevancy.

Sections B.2.1 through B.2.3 present a detailed description of the concept sets searched. Database selection was tailored for each unique set. The specific Dialog® databases searched for each concept set are listed in Sections B.2.1 through B.2.3 by their identifying Dialog® file number designation. In addition, Table B.2

Identifies the database file name with their corresponding file number.

Detailed descriptions of each database are available on the Internet at: <http://library.dialog.com/bluesheets>. The descriptions include information related to subject coverage, date ranges, update frequency, sources of information, record formats, and other aspects.

Information about Dialog® is also available online at: <http://products.dialog.com/products/dialog/index.html>

Table B.2 Dialog® Database Files Searched

File	Name	Date of last update [1] (as of Oct 25, 1999)
2	INSPEC	1969-1999/Oct W1
5	Biosis Previews(R)	1969-1999/Sep W4 [2]
6	NTIS	64-1999/Nov W3via [2]
7	Social SciSearch(R)	1972-1999/Oct W3
8	Ei Compendex(R)	1970-1999/Oct W3
9	Business & Industry(R)	Jul 1994-1999/Oct 25
10	AGRICOLA	70-1999/Oct [2]
14	Mechanical Engineering Abs	1973-1999/Nov
15	ABI/INFORM	Aug 1971-1999
16	Gale Group PROMPT(R)	1990-1999/Oct 25
18	Gale Group F & S Index	1988-1999
19	Chemical Industry Notes (CIN)	1974-1999
20	World Reporter	May 1997-1999
29	Meteor.& Geoastro.Abs.	1970-1999/Sep
31	World Surface Coatings Abs	1976-1999/Jul
33	Aluminum Ind Abs	1968-1999/Nov
34	SciSearch(R) Cited Ref Sc	1990-1999/Oct W3
35	Dissertation Abstracts Online	1861-1999/Oct
40	Enviroline(R)	1975-1999/Jul
41	Pollution Abs	1970-1999/Nov
44	Aquatic Sci&Fish Abs	1978-1999/Oct
47	Gale Group Magazine DB(TM)	1959-1999/Oct 25
49	PAIS INT.	1976-1999/Aug
50	CAB Abstracts	1972-1999/Sep [2]
51	Food Sci.&Tech.Abs	1969-1999/Oct
58	GeoArchive	1974-1999
60	CRIS/USDA	1996-1999
63	Transport Res(TRIS)	1970-1999/Sep
64	Global Mobility Database (R)	1965-1999/Aug
65	Inside Conferences	1993-1999/June W2
67	World Textiles	1970-1999/Sep

Table B.2 Dialog® Database Files Searched (continued)

File	Name	Date of last update [1] (as of Oct 25, 1999)
68	Env.Bib.	1974-1999/Sep
71	ELSEVIER BIOBASE	1994-1999/Sep W2
73	EMBASE	1974-1999/Sep W4
76	Life Sciences Collection	1982-1999/Aug
77	Conference Papers Index	1973-1999
87	TULSA (Petroleum Abs)	1965-1999/Oct W4
89	GeoRef	1785-1999/Sep B2
92	IHS Intl.Stds.& Specs.	1999/Oct
93	TableBase(R)	Sep_1997-1999/Oct W3
94	JICST-EPlus	1985-1999/Jul W1
96	FLUIDEX	1973-1999/Sep
98	General Sci Abs/Full-Text	1984-1999/Sep
99	Wilson Appl. Sci & Tech Abs	1983-1999/Sep
103	Energy Science & Technology	1974-1999
108	Aerospace Database	1962-1999
109	Nuclear Science Abstracts	1948-1976
110	WasteInfo	1974-May/99
111	TGG Natl.Newspaper Index(SM)	1979-1999/Oct 25
115	Research Centers & Services	1994-1998/Dec
116	Brands and Their Companies	—
117	Water Resour.Abs.	1967-1999/Sep
118	ICONDA-Intl Construction	1976-1999/Oct
119	Textile Technol.Dig.	1978-1999/Oct
123	CLAIMS(R)/Current Legal Status	1980-1999/Oct 12
132	S&P`s Daily News	1985-1999/Oct 22
139	Econ. Lit. Index	1969-1999/Oct
141	Readers Guide	1983-1999/Jul
143	Biol. & Agric. Index	1983-1999/Sep [2]
144	Pascal	1973-1999/Sep
146	Washington Post Online	1983-1999/Oct 25
148	Gale Group Trade & Industry DB	1976-1999/Oct 25
155	MEDLINE(R)	1966-1999/Dec W3 [2]
156	Toxline(R)	1965-1999/Sep
161	Occ.Saf.& Hth.	1973-1998/Q3
162	CAB HEALTH	1983-1999/Sep [2]
180	Federal Register	1985-1999/Oct 25
194	CBD	1982/Dec-1999/Jul
195	CBD	Aug 1999-1999/Oct 26
196	FINDEX	1982-1999/Q2
203	AGRIS	1974-1999/Jul [2]
211	Gale Group Newsearch(TM)	1997-1999/Oct 25
238	Abs. in New Tech & Eng.	1981-1999/Oct
240	PAPERCHEM	1967-1999/Jul
241	Elec. Power DB	1972-1999Jan
245	WATERNET(TM)	1971-1999Q1
248	PIRA	1975-1999Nov W4
257	API EnCompass(TM):News	1975-1999/Oct 22

Table B.2 Dialog® Database Files Searched (continued)

File	Name	Date of last update [1] (as of Oct 25, 1999)
258	AP News	Jul_1984-1999/Oct 24
262	CBCA Fulltext	1982-1999/Jul
266	FEDRIP	1999/Jul
269	Materials Bus.(TM)	1985-1999/Nov
285	BioBusiness(R)	1985-1998/Aug W1
286	Biocommerce Abs.& Dir.	1981-1999/Oct B1
292	GEOBASE(TM)	1980-1999/Sep
293	Eng Materials Abs(R)	1986-1999/Nov
315	ChemEng & Biotec Abs	1970-1999/Oct
317	Chemical Safety NewsBase	1981-1999/Oct
318	Chem-Intell Chem Manu Plnts	1999/Jul
319	Chem Bus NewsBase	1984-1999/Oct 25
323	RAPRA Rubber & Plastics	1972-1999/Oct B2
334	Material Safety Label Data	1999/Q2
335	Ceramic Abstracts	1976-1999
337	CHEMTOX (R) Online	1998/Q3
340	CLAIMS(R)/US Patent	1950-99/Oct 12
347	JAPIO - Patent Abstracts of Japan	Oct 1976-1999
348	European Patents	1978-1999/Oct W42
351	DERWENT WPI	1963-1999/ [3]
353	APIPAT	1964-1999/Oct W3
357	Derwent Biotechnology Abs	1982-1999/Sep B1
358	Current BioTech Abs	1983-1999/Sep
430	British Books in Print	1999/Aug
440	Current Contents Search(R)	1990-1999/Oct W5
474	New York Times Abs	1969-1999/Oct 22
475	Wall Street Journal Abs	1973-1999/Oct 22
478	Houston Chronicle	1990-1999/Oct 24
479	Gale Group Company Intelligence(R)	1999/Oct 25
483	NEWSPAPER ABSTRACTS DAILY	1986-1999/Oct 21
484	Periodical Abstracts Plustext	1986-1999/Oct W2
492	Arizona Repub/Phoenix Gaz	1986-1999/Oct 23
494	St LouisPost-Dispatch	1988-1999/Oct 24
495	The Columbus Dispatch	1988-1999/Aug 29
496	The Sacramento Bee	1988-1999/Oct 24
497	(Ft.Lauderdale)Sun-Sentinel	1988-1999/Oct 23
498	Detroit Free Press	1987-1999/Oct 23
515	D&B-Dun`s Elec. Bus. Dir.(TM)	1999/06
516	D & B - Duns Market Identifiers	1999/Aug
519	D&B-Duns Finan.Records Plus(TM)	1999/Apr
527	S&P`s Register-Corp.	1998/Oct
531	Amer. Bus. Directory	1999/Aug
535	Thomas Register Online(R)	1999/Q1
553	Wilson Business Abstracts Full Text	Jan 1983-1999
555	Moody`s(R)Corp.Profiles	1999/Feb W4
559	CORPTECH Dir of Tech Companies	1999/Sep
563	Key Note Market Res.	1986-1999/Oct 24

Table B.2 Dialog® Database Files Searched (continued)

File	Name	Date of last update [1] (as of Oct 25, 1999)
570	Gale Group MARS(R)	1984-1999/Oct 22
571	Piers Exports(US Ports)	1999/Aug
583	Gale Group Globalbase(TM)	1986-1999/Oct 26
584	KOMPASS USA	1999/Jul
603	Newspaper Abstracts	1984-1988
608	KR/T Bus.News.	1992-1999/Oct 13
609	Bridge World Markets News	1989-1999/Oct 24
621	Gale Group New Prod.Annou.(R)	1985-1999/Oct 25
624	McGraw-Hill Publications	1985-1999/Oct 21
630	Los Angeles Times	1993-1999/Oct 23
631	Boston Globe	1980-1999/Oct 22
632	Chicago Tribune	Jan 1988-1999
633	Phil.Inquirer	1983-1999/Oct 24
634	San Jose Mercury	Jun 1985-1999/Oct 16
635	Business Dateline(R)	1985-1999/Oct 22
636	Gale Group Newsletter DB(TM)	1987-1999/Oct 25
637	Journal of Commerce	1986-1999/Oct 22
638	Newsday/New York Newsday	1987-1999/Oct 24
639	The Houston Post	1988-1995/Apr 18
640	San Francisco Chronicle	1988-1999/Oct 23
641	Denver Rky Mtn News	Jun 1989-1999/Oct 22
642	The Charlotte Observer	1988-1999/Oct 24
647	CMP Computer Fulltext	1988-1999
649	Gale Group Newswire ASAP(TM)	1999/Oct 25
652	US Patents Fulltext	1971-1979
653	US Pat.Fulltext	1980-1989
654	US Pat.Full.	1990-1999/Oct 19
655	BNA Daily News from Washington	Jun 1990-1999
660	Federal News Service	1991-1999/Mar 01
665	U.S. Newswire	1995-1999/Apr 29
670	LitAlert	1973-1999/Oct W2
701	St Paul Pioneer Pr Apr	1988-1999/Oct 17
702	Miami Herald	1983-1999/Oct 22
703	USA Today	1989-1999/Oct 22
704	(Portland)The Oregonian	1989-1999/Oct 22
705	The Orlando Sentinel	1988-1999/Oct 24
706	(New Orleans)Times Picayune	1989-1999/Oct 24
707	The Seattle Times	1989-1999/Oct 23
708	Akron Beacon Journal	1989-1999/Oct 24
712	Palm Beach Post	1989-1999/Oct 18
713	Atlanta J/Const.	1989-1999/Oct 25
714	(Baltimore) The Sun	1990-1999/Oct 10
715	Christian Sci.Mon.	1989-1999/Oct 25
716	Daily News Of L.A.	1989-1999/Oct 21
717	The Washington Times	Jun 1989-1999/Oct 22
718	Pittsburgh Post-Gazette	Jun 1990-1999/Oct 22
719	(Albany) The Times Union	Mar 1986-1999/Oct 21

Table B.2 Dialog® Database Files Searched (continued)

File	Name	Date of last update [1] (as of Oct 25, 1999)
720	(Columbia) The State	Dec 1987-1999/Oct 24
721	Lexington Hrld.-Ldr.	1990-1999/Oct 22
723	The Wichita Eagle	1990-1999/Oct 23
724	(Minneapolis)Star Tribune	1989-1996/Feb 04
725	(Cleveland)Plain Dealer	Aug 1991-1999/Oct 23
727	Canadian Newspapers	1990-1999/Oct 24
731	Philad.Dly.News	1983- 1999/Oct 23
732	San Francisco Exam.	1990- 1999/Oct 22
733	The Buffalo News	1990- 1999/Oct 22
734	Dayton Daily News	Oct 1990- 1999/Oct 23
735	St. Petersburg Times	1989- 1999/Oct 23
736	Seattle Post-Int.	1990-1999/Oct 19
737	Anchorage Daily News	1989-1999/Oct 22
738	(Allentown) The Morning Call	1990-1999/Oct 24
739	The Fresno Bee	1990-1999/Oct 23
740	(Memphis)Comm.Appeal	1990-1999/Oct 23
741	(Norfolk)Led./Pil.	1990-1999/Oct 22
742	(Madison)Cap.Tim/Wi.St.J	1990-1999/Oct 23
743	(New Jersey)The Record	1989-1999/Oct 22
747	Newport News Daily Press	1994-1999/Oct 24
763	Freedonia Market Res.	1990-1999/Jul
764	BCC Market Research	1989-1999/Sep
765	Frost & Sullivan	1992-1999/Apr
766	(R)Kalorama Info Market Res.	1993-1999/Sep
777	EdgarPlus(TM)-Annual Reports	1999/Oct 22
781	ProQuest Newsstand	1998-1999/Oct 24
813	PR Newswire	1987-1999/Apr 30
861	UPI News	1996-1999/May 27

[1] W = week, B = biweekly, M = month, Q = quarter

[2] This file was also reviewed in its CD format.

[3] Note: UD=, UM=, & UP=199943

B.4 External Review

This section describes the external review process used in Phase 1. Instructions to the reviewers is outlined in Section B.4.2. Section B.4.3 outlines actions taken in response to the reviewer comments.

B.4.1 Introduction

External review of project results by non-NAL experts was a key element of the QA/QC Plan.

As a result the NAL staff recruited qualified independent reviewers in the fields of soil science, civil engineering, and information science. It should be noted that the reviews were voluntarily conducted without compensation.

B.4.2 Reviewer Instructions

NAL requested that each reviewer consider four specific elements in conducting the review of project results:

(1) Identify additional terms.

(2) Review search string syntax for logical construction.

(3) Review database selection.

(4) Identify any known highly relevant sources not presented in the reviewer package.

B.4.3 Summary of Actions Taken in Response to the External Reviewers' Comments

The responses of external reviewers for this information research project included comments, suggestions, and minor corrections to improve the searches and the information subsequently retrieved. All reviewer notes were addressed in revisions or additions to the comprehensive strategy statements (Section B.2), or through additional Web searches, and the inclusion of pages from noted sites for examination by the NRC staff. These changes were provided to the NRC in the "Supplement to the Draft Letter Report," dated October 13, 1999 and the "Final Letter Report" dated November 1999.

Carol Reese of the American Society of Civil Engineering (ASCE) suggested three specific changes and additions to the basic search strategy text. Specifically, these included using "removal" in addition to "remove" in concept set V3. Note that the final executed version of this command line includes "remov?", shown in bold face text below. This truncated form retrieves all endings for the "remov" root, so that remove, removal, removing, removed, etc., were all included and retrieved in the final searches for this set.

Concept Set V3:
S (SOIL? ? OR DIRT) /TI,DE,ID,SH,CC (F) (TRUCK? OR SHIP? OR TRANSPORT? OR HAUL? OR BARG? OR TRAIN? OR RAIL? OR CONVEY??? OR **REMOV?** OR RELOCAT? OR REPLAC? OR PLACE? ? OR PLACEMENT)

Ms. Reese also noted that the term "cost" had been omitted from the draft version of the search concept set labeled V1. This change was also completed before final execution of the

comprehensive search statements, using the truncated form "cost???" (shown below in bold face type). The final version of this set retrieved all
endings with up to three characters following the "cost" root, and included cost, costs, costing, costed.

Concept Set V1
S (SOIL? ? OR DIRT) /TI,DE,ID,SH,CC (S) (WHOLESALE? OR PROFIT? OR **COST???** OR INDUSTR??? OR COMMERC??? OR BUSINESS?? OR INVEST? OR MARKET??? OR SALE? ? OR PURCHAS??? OR DOLLAR? ?)

In addition, Ms. Reese observed that a closing parenthesis was needed in the P3 concept set. This error in the draft version was noted and corrected in execution of the final searches, as shown in boldface type below.

Concept Set P3:
S (SOIL? ? OR DIRT OR EARTH? ?**)** /TI,DE,ID,SH,CC

Ms. Reese also recommended two additional information providers, STN and CISTI: however, both were examined and determined not to be significant new sources of project-relevant information. Considerable overlap of database coverage exists between the STN and Dialog® database systems. The notable strengths in the STN database system include intellectual property and patents, materials and mechanical engineering, and German-language sources. Each of these categories had previously been excluded by mutual agreement. Therefore, NAL researchers concluded that searching STN was unlikely to yield new or unique results in any meaningful quantity.

CISTI, is primarily a publicizing and document ordering service available over the Internet at: <http://cat.cisti.nrc.ca>. Survey searches conducted in the CISTI system indicated fewer than 7,500 records containing the terms "soil" or "soils" in the combined catalogues. Without the availability of a sophisticated search engine on the site, in-depth research of this resource was not deemed to be cost- effective. The selected list of

CISTI indexed journals suggested by Ms. Reese were in each case indexed in multiple DIALOG® files searched for this study.

Dr. Dewayne Mays of the USDA Soil Survey Laboratory did not comment on the strategy, but he did suggest specific data sources that should be used for the searches, including theses and dissertations. Dialog® file 35, "Dissertation Abstracts," contains these document types. This database was one of the key files searched for the comprehensive titles listing delivered to the NRC on September 29, 1999, as listed in Appendix A. This database file was included in 15 of the 42 searches that produced the comprehensive lists of titles delivered to the NRC. At least one thesis title was initially selected by the NRC.

Dr. Mays also noted a specific Web site covering USGS documents. The USGS database was among those included in the comprehensive searches conducted on the Dialog® system, listed in Table B.2 as file 89. This Dialog® database file was used in 25 of the 42 specific searches completed.

Ms. Carla Casler, of the Arid Lands Research Institute, added no specific comments on the strategy statements, but noted the need to consider international sources in the information survey and review processes. Although search results were limited to English language documents, a great many of the databases searched included international literature by default because a vast majority of these items are published in English or are posted with English language titles. This enabled their retrieval despite the use of English language limitations. A number of these non-English records were selected by the NRC staff.

In addition, Ms. Casler suggested the following Web sites for access to specific reports covering radiation exposure incidents at Three Mile Island and the Chernobyl site:

- <http://www.iaea.or.at/worldatom/thisweek/preview/chernobyl/ paper5.html>
- <http://www.libraries.psu.edu/crsweb/tmi/resources.htm>

Pages from both sites were forwarded to the NRC staff for their review. Note that the NRC staff was familiar with the Three Mile Island Web materials.

Ms. Casler made another point in reference to the long-term and epidemiological impacts of radiation exposure through soil contact scenarios of various sorts, including gardening. Specifically, she noted that it might be valuable to consider Russian literature covering Chernobyl, as well as sources about the long-term effects or impacts of the Hiroshima and Nagasaki bombs on soils.

Studies related to these topics were among the many items selected by the NRC staff from the titles listed in the text of the draft letter report and the files of titles in the comprehensive search results. Note that the unintentional exposure hazard from high-level radiation that occurred in the cases Ms. Casler mentioned is significantly different from the anticipated exposure derived from soils intentionally released from the NRC-regulated locations.

B.5 Selected Internet Resources From Phase 1

The primary focus for this study was to identify verifiable information by conducting traditional literature research. Nonetheless, in today's information environment, no research project can be considered complete without a preliminary survey of the Internet.

The Internet is an important new source of information. However, as valuable as this resource undoubtedly is, it has significant limitations, most notable among which is the inability to conduct comprehensive complex searches.

Search engines and specific resources searched are described in Section 4 of this report. Notable information resources were discovered. These resources were reviewed by the NRC staff, who selected the following:

- Ohio Department of Natural Resources, Geological Survey, "GeoFacts No. 19, Sand and Gravel," <http://www.dnr.state.oh.us/odnr/ geo_survey/geo_fact/geo_f19/ geo_f19.htm>

- Brookhaven National Laboratory, Upton, New York, "Important Web Links," <http://www.dne.bnl.gov/ssn/ Weblinks.html>

- Ohio Site Technology Coordination Group, "Technology Needs, Ashtabula," <http://www.ohio.doe.gov/oh-stcg/ needs.asp>

- Mineralogical Society, "Publications," <http://www.minersoc.org/ publicat.htm>

- Clay Minerals Society, "Homepage," <http://cms.lanl.gov/>

- US Mix, "US Mix Products," <http://www.usmix.com/usmix.html>

- Oklahoma State University, Plant and Soil Sciences Department, "CMLS94: Chemical Movement in Layered Soils," <http://clay.agr.okstate.edu/software/cmls94a. htm>

- Canadian Government, Department of Indian and Northern Affairs, "Building a Future, Sand and Gravel," <http://www.inac.gc.ca/building/ sands/sand.html>

- Off-Road.com, "DirtBikes Online," <http://www.off-road.com/dirtbike/>

- National Dirt Digest, "Dirt Late Model News," <http://www.latemodel.com/ nddigest/>

- AMA Pro Racing, "Dirt Track," <http://www.ama-cycle.org/prorace/99dt/99dt. html>

- Bolin Enterprise, Inc., PowerLift Foundation, Repair Division, "Foundation Repair Specialists," <http://www.foundationspecialists. com/html/advanced.htm>

- A.B. Chance, Hubbell Power Systems, "Earth Anchors and Foundations," <http://www.hubbell.com/ abchance/>

- ENA, Inc., "Excavation and Road Construction Specialists," <http://a1.com/ena/index.html>

- PRISM - World Resource Foundation, "Landfill Mining Technical Brief," <http://www.wrfound.org.uk/ wrftblfm.html>

- Purdue News, "Purdue-Made Soil," <http://www.purdue.edu/UNS/html4ever/97060 6.Tishmack.soil.html>

- McGraw-Hill Construction Information Group, "Sweets Web Links," <http://www.sweets.com/topic/ weblinks.htm>

- New Jersey, Department of Environmental Protection, "Site Remediation Program 1998 Revised Guidance Document for the Remediation of Contaminated Soils," <http://www.state.nj.us/dep/srp/regs/ soilguide/>

- CRC Press LLC Online, "The Reuse & Recycling of Contaminated Soil," <http://www.crcpress.com/index.htm?catalog/ L1188>

- Appropriate Technology Transfer for Rural Areas (ATTRA), "Organic Potting Mixes - Horticulture Technical Notes," <http://www.attra.org/attra-pub/ potmix.html>.

APPENDIX C. PHASE 2 ADDITIONAL RESOURCES

C.1 Information Resources for the Commercial Sector

Additional information resources for the commercial sector include inventory tracking firms, industry and professional associations, and market survey reports and firms as discussed in Sections C.1.1 through C.1.3.

C.1.1 Inventory Tracking Firms

SKUfinder.com and **1-800-DATABASE, LTD.** work in partnership with product manufacturers and their trade associations. The company has a comprehensive database containing images and information about hundreds of thousands of consumer products, including information from the American Hardware Manufacturers Association. (See <http://www.skufinder.com/>.)

Vista Information Services produces market research reports on the basis of point-of sale scanner data. These reports track products that are sold in key home improvement retail channels, and include information about independent hardware stores, lumber yards, home centers (Home Depot, Lowes, etc.) and mass merchants. (See <http://www.triad.com/vista/home.htm>.)

C.1.2 Industry and Professional Associations

TechSavvy.com provides company directories, historical data, commercial and industry standards, U.S. military documents and parts information. Some of the identified titles are " Specifications for topsoil used for landscaping purposes," "Specification for topsoil," and "soils manuals." (See <http://www.techsavvy.com>.)

The following organizations are identified as possible sources of information on soil which may be related to soil reuse:

- Mulch and Soil Council (formerly the National Bark and Soil Producers Association), 10210 Leatherleaf Ct., Manassas, VA 20111-4245, WEB: <http://www.nbspa.org/>

- Lawn & Garden Marketing and Distribution Association, 1900 Arch Street, Philadelphia, PA 19103-1498, telephone: 215-564-3484, WEB: <http://www.lgmda.org/>

- National Gardening Association, 1100 Dorset St., South Burlington, VT 05401, telephone: 802- 863-5251, WEB: <http://www.garden.org/>

- American Nursery & Landscape Association, 1250 I Street, NW Suite 500, Washington, DC 20005-3922, telephone: 202-789-2900, WEB: <http://www.anla.org/>

- Materials Handling & Management Society, 8720 Red Oak Blvd., Suite 201, Charlotte, NC 28217, telephone: 704-676-1183, WEB: <http://www.mhia.org/ps/PS_MHMS_Home.cfm>

- Warehousing Education and Research Council, 1100 Jorie Blvd., Suite 170, Oak Brook, IL 60523-4413, telephone: 630-990-0001, WEB: <http://www.werc.org/>

C.1.3 Marketing Survey Reports and Firms

A review of the marketing survey reports and firms identified the following related groups and reports:

- The Annual National Garden Survey, sponsored by the National Gardening Association

- Lawn and Garden Market, report by Package Facts

- Kline and Co. Survey of the Lawn and Garden Industry

- American Community Gardening Association

- Canadian Office of Urban Agriculture.

C.2 Climate and Crops

Soil use varies with respect to climate and weather. Information on these factors, on a national scale, is available via several Internet resources. In particular, the National Atlas of the United States of America, available at <http://www.nationalatlas.gov> provides maps showing first and last freeze dates in autumn and spring. Temperature means, maxima, and minima are also shown. In addition, maps are available that graphically depict the winds, average rainfall, days of precipitation and sunshine, and sky cover data for the United States.

The USDA's Agriculture Handbook #664, "Major Crop Areas and Climatic Profiles," provides a discussion of crop-climate associations. Other online USDA resources offer additional information of this type:

- mean annual precipitation maps (with access to State-specific maps, and data sets) available at <http://www.ftw.nrcs.usda.gov/prism/prism.html> **or**

<http://www.wcc.nrcs.usda.gov/water/climate/prism/prism.html>

- National and State-specific graphics and data for precipitation and temperature, available at <http://www.ftw.nrcs.usda.gov/prism/prismmaps.html>.

In addition, the NASS provides Web access to graphics showing the kinds of data described above. NASS also provides crop-specific planting maps, data, and weather details; labor; production expenditures; land in farms; and other graphic data sets at National and State-specific levels available at <http://www.usda.gov/nass/aggraphs/graphics.htm>.

Long-term plant survival is dependent on a number of factors including the ability to survive low temperatures over the winter. In 1960 USDA/Agricultural Research Service scientists at the U.S. National Arboretum collaborated with the American Horticultural Society and other horticulturists to create the USDA "Plant Hardiness Zone Map." Map zones are based on validated average annual minimum temperatures. Average temperatures were determined using at least 10 years of data. The most up-to-date version of the USDA "Hardiness Zone Map" is available at <http://www.usna.usda.gov/Hardzone/hrdzon2.html>.

APPENDIX D. GENERAL INFORMATION ON SOILS

D.1 The Nature of Soils

Soils comprise mixtures of mineral and biotic components, with spaces where air or water are found. Soils are often characterized according to the ratio of three principle components, namely sand, silt, and clay (in order of decreasing particle size). Different soils develop in different locations, as a result of the particular impacts of local environmental factors with specific local geologic (parent) materials. They also develop and change over time through the interactions of the parent materials with the local topography, weather and climate, vegetation, and external disturbances.

Soils provide physical support for plants, as well as their supply of nutrient ions. Available nutrients from the soil are delivered to plants via water from layers within the soils; water in the soil pore spaces and in the porous materials of the soil; or water bound in clay mineral layers. Soil air spaces allow gas exchange to the plant roots. Through these processes, soils enable the production of food, animal feed, and fiber for the world's populations.

Soils also provide structural support for buildings and insulation for inhabited areas that are underground. Soils may be used as part of the foundation for roads, highways and other construction. Soils also provide a habitat for large and small animals, insects and microbes.

D.2 Soil Variability

Soils vary greatly over regions, within local areas, or even within a single farm field, lawn, or garden plot. Variations may include differences in texture (relative proportions of sand, silt and clay particle sizes), structure (size and shape of soil aggregates or blocks), moisture characteristics (water content, water supply capacity and water-holding capacity), chemistry and reactivity (amounts, kinds, and forms of minerals, elements, and ions) and biota (soil-dwelling microbial, insect, and other arthropod species). Soil biota may include fungi, bacteria, viruses, nematodes, worms, and others—both vegetative bodies and individuals, as well as propagules (eggs, spores, cysts, etc.) or virus particles, fragments, etc., and wastes and residues from once-living organisms.

Information on soil and soil types, as well as the dominant soil orders (major groups of soil types) and their distribution may be seen online at <http://www.nhq.nrcs.usda. gov/land/index/soils.html>. The characteristics and properties used to define different soils in the United States are presented in the USDA's Agriculture Handbook #436 "Soil Taxonomy: A Basic System of Soil Classification for Making and Interpreting Soil Surveys," (second edition 1999) prepared by the Natural Resources Conservation Service (NRCS) and is available at <http://www.statlab.iastate.edu/ soils/soiltax/>. Similarly, soil taxonomy maps are also provided in USDA, 1999. For the latest map of dominant soil orders in the United States, derived from the State Soil Geographic Database (STATSGO), see <http://www.nhq.nrcs.usda.gov/land/ meta/m4025.html>.

Soil scientists in the United States and worldwide have made considerable effort to study, identify, and classify soils for agriculture, construction, recreation and other uses. The USDA's Soil Survey program, in operation for more than 100 years, has described and classified the soils in nearly every region of the United States, on a county-by-county basis.

Access to soil survey work is available from the Web site of the USDA's NRCS (previously known as the Soil Conservation Service) found at <http://www.statlab.iastate.edu/soils/soildiv/sslists /sslisthome.html>

From this Web site, one may obtain a recently updated survey for each State by accessing the State-specific pages, which are accessed from <http://www.statlab.iastate.edu/soils/soildiv/survey

s/onlineman.html>.

Printed soil surveys are available in local (county) libraries, or from the land-grant university library in each State. A list of these locations may be obtained from <http://www.nal.usda.gov/pubs_dbs/ landgrant.htm >.

Other USDA NRCS Web pages provide considerable information about soils such as the "Keys to Soil Taxonomy" available at <http://www.statlab.iastate.edu/soils/ keytax/>. Additional valuable online information resources are available at the National Soil Survey Center Soil Science Education Web site at <http://www.statlab. iastate.edu/soils/nssc/ educ/Edpage.html>.

D.2.1   Soil Information Resources

In addition to NRCS resources, the Internet offers many other useful Web sites for soil-related information. These may be helpful in learning more about soils in general, or about specific soil studies:

- "Soil Science Literature Searching— General Soil Science" (from Humboldt University) <http://library.humboldt.edu/~rls/ gensoil.htm>

- "Soil Science Resources on the Internet" (from the University of Wisconsin) <http://www.library.wisc.edu/libraries/ Steenbock/electron/soilsci.htm>

- "Soil and Water Web" (Dr. John E. Thomas, Gainesville FL) <http://soilweb.tripod.com/ soilsite.htm>.

D.2.2   Soil and Farming/Land Use

Soil suitability for particular crops or farming and production practices is detailed extensively in soil survey reports. The USDA Land Use Capability Classification scheme comprises eight major classes (Class I–Class VIII). Only Classes I–IV are suited for cultivation, and Classes III and IV

have severe limitations to crop production that must be overcome by limiting crops and practices, and/or extensive management input. In addition to limits indicated by land use capability classification of specific soils, other limits on soil use are determined by weather and climate, crop tolerances (for weather, wetness, salinity, etc.), and length of the growing season.

Information on specific agricultural uses of soils in the Untied States is provided by a number of online and printed resources. A few of these are Web sites; some are also available in printed form:

- "Table 4— Land Cover/Use on NonFederal Rural Land, by Land Capability Class and Subclass, by Year," from the 1997 National Resources Inventory—Summary Report" <http://www.nhq.nrcs.usda.gov /NRI/1997/summary_report/original /table4.html>

- "Agricultural Resources and Environmental Indicators, 2000," (including figures and tables with specific land use details) from USDA's Economic Research Service, Resource Economics Division <http://www.ers.usda.gov/emphases/harmony /issues/arei2000/AREI1_1landuse.pdf>

- "Percent of Non-Federal Area in Cultivated Cropland, 1997," from NRCS's Resource Assessment Division <http://www.nhq.nrcs.usda.gov/land/meta/m4 962.html>

- "Figure 1. How Our Land is Used," from the 1997 National Resources Inventory—Summary Report <http://www.nhq.nrcs.usda.gov/NRI/1997/sum mary_report/original/figures.html#figure1>

- "Broad Land Cover/Use by State, 1997" from the NRCS <http://www.nhq.nrcs.usda.gov/land/lgif/ m5150l.gif>.

Other graphic displays covering land use are available from USDA NRCS, Web pages at <http://www.nhq.nrcs.usda.gov/ land/index/>. Crop specific information is also available from <http://www.nhq.nrcs.usda.gov/land/index/croplnd .html>.

D.3 Additional Information on Soil

Because soils are heavy and variable, their use is largely dictated by location. Soil materials are heavy because they have high particle densities (averaging about 2.6 g/cc) and a bulk density (including soil pore spaces) that averages about 1.3 g/cc for cropped mineral soils. This means that an acre (43,560 square feet) of field soil (dry) 1 foot deep would weigh more than 3.5 million pounds (about 1,766 tons). The ratio of bulk density to particle density represents the pore space in the soil, which is normally filled with air and/or water. If the same soil was fully wetted, the weight of that acre-foot of soil would increase by another 1.3 million pounds. With a truck shipment cost of around $0.26 per ton-mile, the cost of moving 1 acre of soil 1 foot deep would be over $450 for each mile traveled, thus generally prohibiting the relocation of soil for crop production. Calculations and cost data are shown at <http://www.soils.umn.edu/academics/ classes/soil3125/doc/lecslwt.htm> and <http://www.hq.usace.army.mil/cecw/fusrap/techp ap/pkgin.htm>.

The use of natural "soil" in horticulture has continued to decline since the publication of the early work on using artificial planting media (e.g., the U.C. mix) for the production of horticultural crops, vegetables, fruits and ornamental potted (i.e, containerized) plants, transplants, and greenhouse plugs (Baker, 1957). Lightweight non-soil materials are now the standard substrate for modern nursery and greenhouse production. In addition to peat moss, pearlite, and vermiculite, numerous materials from agricultural or industrial wastes and byproducts have been found to be very helpful for horticultural production applications. Additionally, new media materials (such as foams, gels and polymers) have been developed and tested, many are now in use.

There are many reasons to avoid soil use in containerized planting. Bagged materials containing soil are heavier to prepare, process and deliver than other planting mixes. Soil use in horticultural nursery and greenhouse production, limited by weight and transportation costs, is further minimized by quality and consistency requirements. Loss of water flow pathways alter a soil's water supply characteristics in greenhouse or other potted plant applications. Also because natural soils have biotic components, they must be sterilized or treated to remove pathogens, insects, weed seeds, etc. In addition, the grower's needs to regulate and control fertility, pH, and water supply capacity are more readily met by using defined and tested mixes of non-soil media materials. With defined media, nutrient additions may be more readily calibrated and controlled to supply needed fertility levels for each specific application.

Limited use of soils or soil materials does occur in particular situations, especially where the added water supply capacity of soils or soil components is needed, as in tropical nurseries. For example, soil was once the basis of most potting mixes. It is still used in nurseries in the tropics as a means of increasing water-holding capacity of media for large pots. Elsewhere, most media are soil-less (Handreck, 1994). Growers can sometimes increase the water supply capacity of growing media by adding clay to the potting mix (usually a particular clay mineral), rather than adding a composite (natural) soil material.

Many growing media mixes and similar products offered by nursery and garden supply companies, and also used by commercial growers, may be named with phrases such as "potting soil", or "topsoil". Nonetheless, only a few of these products actually contain any soil at all. Inconsistencies among even similar soils, the need for fumigation or sterilization to preclude disease and pest infestation, and the weight and related transport costs make it desirable for both nursery producers and suppliers to use standard media mixes, with predictable characteristics and performance for their production needs.

Informal surveys in local retail outlets in Queen Anne's County on the Eastern Shore of Maryland revealed few examples of potting mix or "topsoil" products that contained any soil at all. Ingredient lists from some products labeled "topsoil" did include "soil" materials, but often only as a second or third ingredient. Media product label information and packaging text examples (from product lists) were located at <http://www.hortnet.com> and <http://www.vitalearth.com>.

Soil transfer or removal and relocation in construction may involve moving soil from one area to another, but soils rarely shipped more than short distances, because of the costs. One notable exception to this might be the case of reclamation processing of contaminated soils, from chemical waste contamination or spills. In many such cases, when a large amount of soil is being processed, the machinery may be brought to the site of the contaminated soil and treated *in situ*.

APPENDIX E. NAL PROJECT INVESTIGATORS

The U.S. Nuclear Regulatory Commission (NRC) and the National Agricultural Library (NAL) established an interagency agreement to conduct information research on human-soil interactions. This appendix identifies the NAL investigators who participated in this study.

- **Abiola Adeyemi, B.S.**
  (Phase 1 team member)
  Program Assistant/Information Specialist
  Alternative Farming Systems Information Center

Mr. Adeyemi holds a Bachelor of Science degree from the University of Maryland Department of Agriculture and Natural Resources. He also has extensive training in international agroecology from the University of California, Berkeley. He specializes in information gathering in the areas of sustainable agriculture, urban/community farming, and international development.

- **Tim Allen, M.S.**
  (Phase 1 team member)
  Reference Consultant and Technical Information Specialist
  Animal Welfare Information Center

Mr. Allen has a Master of Science in Animal Science. He conducted research in private industry before joining the National Agricultural Library. Mr. Allen has extensive search and retrieval experience and conducted the first Defense Technical Information Center Web search. In addition, he served as an internal project reviewer.

- **Andy Clark, Ph.D.**
  (Phase 1 team member)
  Coordinator, Sustainable Agriculture Network
  Reference Consultant

Dr. Clark has a Ph.D. in Agronomy and serves as the Coordinator for the Sustainable Agriculture Network, a part of the USDA-funded Sustainable

Agricultural Research and Education Program. For this study, he helped screen the initial large G1 data set, and his selections were incorporated into the draft letter report.

- **Stuart Gagnon, M.S.L.S.**
  (Phase 2 team member)
  Librarian
  Water Quality Information Center

Mr. Gagnon graduated from the University of North Carolina at Chapel Hill's School of Information and Library Science (SILS) with a Master of Science degree in Library Science. He has worked in environmental librarianship (developing reference resources, building Web-accessible databases and researching environmental terminology) for several years. Mr. Gagnon is a cooperator with the University of Maryland Libraries.

- **Mary V. Gold, M.L.S.**
  (Phase 2 team member)
  Librarian/Information Specialist
  Alternative Farming Systems Information Center (AFSIC)

Ms. Gold has a Master of Science degree in Library Science and, for the past several years, has specialized in reference services related to alternative and sustainable agriculture systems, including the searching and retrieval of electronic and print information. She has also been trained as a Master Gardener with the University of Maryland Cooperative Extension Service, and is a member of the Garden Writers Association of America.

- **Terrance Henrichs**
  (Phase 1 and 2 team member)
  Program Support Assistant.

Ms. Henrichs led the project support effort in formatting, sorting, and compiling the initial bibliography of both the Phase 1 and 2 final letter reports. She also composed Figure 1, and provided other graphical support. In addition, she provided support in copying and shipping reports to the NRC.

- **Judy Keen, M.L.S.**
  (Phase 2 team member)
  Librarian and Reference Specialist Information Research Services Branch

Ms. Keen earned a Master of Library Sciences degree from Catholic University. She is a key reference specialist located at NAL's Washington D.C. Reference Center. In addition to her work with the NAL Phase 2 Soil Working Group, Ms. Keen has recently worked with teams working on invasive species and "Biosecurity, the Threat to Agriculture."

- **Susan McCarthy, Ph.D.**
  (Phase 1 and 2 team member)
  Technical Information Specialist
  Principal Investigator and coauthor

Dr. McCarthy has a Ph.D. in Plant Physiology, 9 years of experience in reference services, and more than 15 years of bench research. She conducted Internet and CD-ROM searches for the project, coauthored the draft, supplement, and final reports, and served as project manager.

- **Sharon Middleton**
  (Phase 1 team member)
  Program Assistant

Ms. Middleton provided project support in report assembly and packaging, and shipped copies of the reports to the external reviewers and NRC staff. Her most important role was to reformat selected citations for the Phase 1 final letter report.

- **Maria Pisa, M.L.S.**
  (Phase 1 and 2 team member)
  Associate Director of Public Services
  Principal Investigator for administrative issues and coauthor of draft report

Ms. Pisa has a Master of Library Science degree

and more than 20 years experience in library services.

- **M. Louise Reynnells, B.S.**
  (Phase 2 team member)
  Technical Information Specialist
  Rural Information Center

Mrs. Reynnells has 14 years of experience as a subject specialist in rural agricultural and community development issues. Prior to her current position, Mrs. Reynnells worked at the University of California, Cooperative Extension Service in Riverside, CA, as a State Research Associate doing original research in the poultry sciences. Mrs. Reynnells' educational background includes a Bachelors degree in Agricultural Science/Animal Science, from California State Polytechnic University, which included course work in soil sciences. In addition, her background includes graduate work in Library and Information Sciences at the University of Maryland.

- **Karl Schneider, M.L.S.**
  (Phase 1 and 2 team member)
  Reference Specialist in Soils
  Principal Investigator for searching and subject matter and coauthor

Mr. Schneider has a Master of Library Science degree, many years of reference and online search experience, as well as graduate research and training in soils and related sciences.

- **Michael Shochet, M.L.S.**
  (Phase 2 team member)
  Reference Librarian
  Technology Transfer Information Center

Mr. Shochet provides scientific, technical, and business information regarding patents, and the commercialization of new technologies to an international community. He earned his Bachelor of Arts and Master of Arts degrees in Anthropology from Cornell University and Northwestern University, respectively, and his Master of Library Science degree from the University of Maryland.

His previous experiences include providing reference services to support Navigant Consulting, a business firm that specializes in valuations of intellectual property; providing reference and bibliographic instruction services at the Baltimore City Community College; and assisting people with disabilities to develop and achieve their goals and objectives.

- **Mary Stevanus, M.L.S.**
  (Phase 1 Team Member)
  Reference Specialist

A principal contributor to the Phase 1 draft letter report.

Ms. Stevanus has many years of reference and online search experience. She also worked for a number of years as an information specialist for the U.S. Environmental Protection Agency. In Phase 1 of this study, Ms. Stevanus served as an internal project reviewer, refined the comprehensive strategies used in the draft letter report, and conducted WorldCat, Internet, and other searches.

NRC FORM 335
(2-89)
NRCM 1102,
3201, 3202

U.S. NUCLEAR REGULATORY COMMISSION

BIBLIOGRAPHIC DATA SHEET

(See instructions on the reverse)

1. REPORT NUMBER
(Assigned by NRC, Add Vol., Supp., Rev.,
and Addendum Numbers, if any.)

NUREG-1725

2. TITLE AND SUBTITLE

Human Interaction with Reused Soil: An Information Search

Final Report

3. DATE REPORT PUBLISHED

MONTH	YEAR
January	2002

4. FIN OR GRANT NUMBER

Y6227

5. AUTHOR(S)

S. McCarthy, USDA/ARS/NAL; T. Nicholson and J. Philip, USNRC/RES; E. Brummett,
F. Cardile, G. Gnugnoli and A. Huffert, USNRC/NMSS

6. TYPE OF REPORT

Technical

7. PERIOD COVERED *(Inclusive Dates)*

August 1999 - January 2002

8. PERFORMING ORGANIZATION - NAME AND ADDRESS *(If NRC, provide Division, Office or Region, U.S. Nuclear Regulatory Commission, and mailing address; if contractor, provide name and mailing address.)*

U.S. Department of Agriculture

National Agricultural Library

10301 Baltimore Avenue

Beltsville, MD 20705-2351

9. SPONSORING ORGANIZATION - NAME AND ADDRESS *(If NRC, type "Same as above"; if contractor, provide NRC Division, Office or Region, U.S. Nuclear Regulatory Commission, and mailing address.)*

Division of Systems Analysis and Regulatory Effectiveness

Office of Nuclear Regulatory Research

U.S. Nuclear Regulatory Commission

Washington, DC 20555-0001

10. SUPPLEMENTARY NOTES

T.J. Nicholson, NRC Project Manager

11. ABSTRACT *(200 words or less)*

This NUREG-series publication reports the results of a 2-year investigation to compile information intended to support the formulation and characterization of scenarios related to exposure to residual radioactivity in reused soils. This information search focused on human interactions with reused soils in the United States. Using this information, the staff and contractors of the U.S. Nuclear Regulatory Commission (NRC) are working to define realistic soil reuse scenarios and to estimate parameters for simulating exposure pathways if soil is removed from NRC-licensed facilities. NRC staff and researchers from the National Agricultural Library (NAL) conducted this investigation in two phases. Phase 1 was a general information search structured to query literature from a wide range of published scientific and trade sources. Phase 2 was a focused information search on specific parameters such as contact time, dust exposures and tillage depths identified in Phase 1. NAL staff searched additional sources and contacted individuals in the Government, academia, and commerce. This report compiles, for the first time, data and information sources for parameters specific to soil reuse. This report also provides information that is relevant for generic, as well as site-specific dose assessments, and presents typical information that may be used in future dose modeling analyses. This investigation was coordinated with a companion effort by the Federal Interagency Steering Committee on Radiation Standards on disposition of sewage sludge.

12. KEY WORDS/DESCRIPTORS *(List words or phrases that will assist researchers in locating the report.)*

clearance of soil
dose assessment scenarios
human interaction with reused soil
information search strategy
recycled soil
residual radioactivity
reused soil
soil removed from NRC-licensed facilities
soil reuse scenarios

13. AVAILABILITY STATEMENT

unlimited

14. SECURITY CLASSIFICATION

(This Page)

unclassified

(This Report)

unclassified

15. NUMBER OF PAGES

16. PRICE

NRC FORM 335 (2-89)

This form was electronically produced by Elite Federal Forms, Inc.

www.ingramcontent.com/pod-product-compliance
Lightning Source LLC
Chambersburg PA
CBHW081457170526
45166CB00008B/2461